SCULPTED by ICE

Glaciers and the Alaska Landscape

SCULPTED *by* ICE
Glaciers and the Alaska Landscape

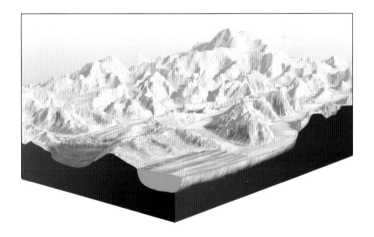

by Michael Collier

Alaska Natural History Association
Anchorage, Alaska

DEDICATED TO KEITH ECHELMEYER—
AN EXCEPTIONAL SCIENTIST AND AN EXTRAORDINARY HUMAN BEING

Author: Michael Collier
Designer: Carole Thickstun, Ormsby & Thickstun Design
Illustrations/Maps: Lawrence Ormsby, Ormsby & Thickstun Design
Editor: Susan Tasaki
Project Coordinator: Lisa Oakley

Contemporary Photographs: Except as noted, all photographs are
© Michael Collier. pp. 80-81, Burnie Schultz: p. 105, Cynthia
D'Vincent

Historical Photographs: p. 25, William O. Field, compiler. 1978;
pp. 28-29, Austin S. Post. Glacier Photograph Collection. Boulder,
CO: National Snow and Ice Data Center

Publication supported by a grant from the
U.S. Geological Survey

750 West Second Avenue, Suite 100
Anchorage, AK 99501
www.alaskanha.org

Alaska Natural History Association is a nonprofit educational partner
of Alaska's parks, forests and refuges. In addition to publishing books
and other materials about Alaska's public lands, Alaska Natural
History offers field-based educational programs, teacher trainings and
operates visitor center bookstores. The net proceeds from publication
sales support educational programs that connect people to the natural
and cultural heritage of Alaska's public lands. For more information
or to become a supporting member: www.alaskanha.org

Library of Congress Cataloging-in-Publication Data
Collier, Michael, 1950-
Sculpted by ice: glaciers and the Alaska landscape/text and photo-
graphs by Michael Collier; illustrations by Lawrence Ormsby
p. cm.
Includes bibliographical references
ISBN 0-930931-23-8 (pbk.)
1. Glaciers-Alaska. 2. Landforms-Alaska I. Title: Glaciers and the
Alaskan landscape. II. Title.

GB2425.A4C65 2004
551.31'2'09798--dcc22
2004001467

Printed in China on recycled paper using soy-based inks.

p. i: *Colony Glacier, east of Palmer, Chugach National Forest.*
p. ii: *Ridge above Muldrow Glacier, Denali National Park and Preserve.*
p. v: *Herbert Glacier, northwest of Juneau, Tongass National Forest.*
p. vi: *John Hopkins Glacier, Glacier Bay National Park and Preserve.*

CONTENTS

Chapter One

SEEING IS BELIEVING

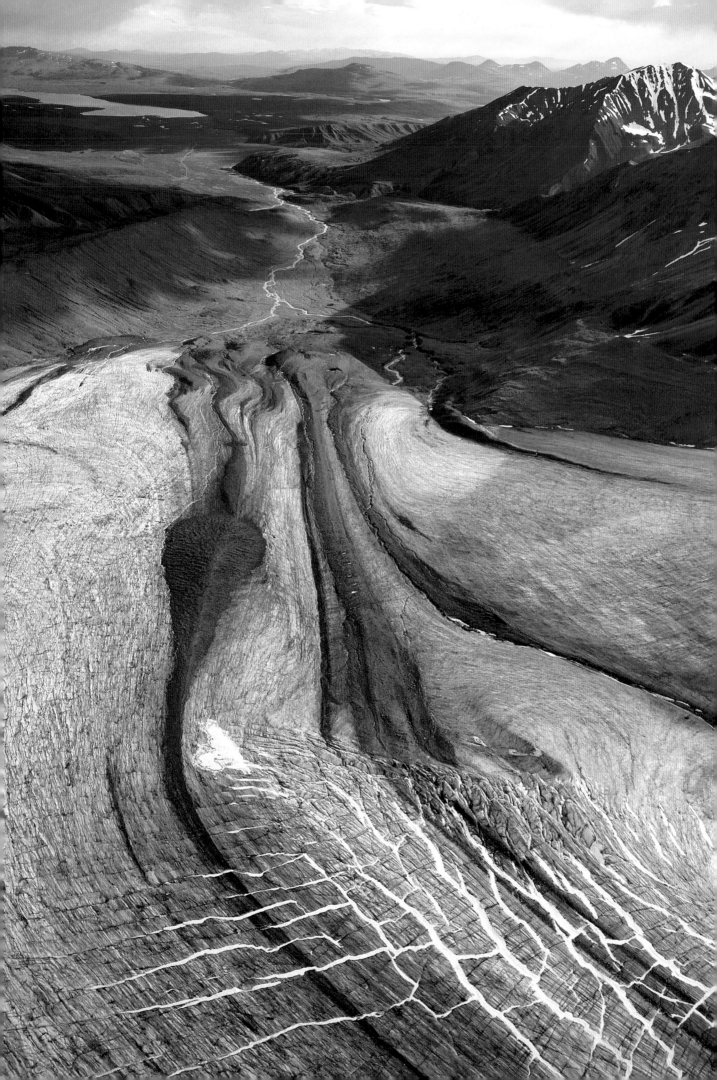

A storm was building. Fitful gusts of wind stirred the snowdrifts above us. We had worked our way around Gulkana Glacier's terminus, up its west flank, under an icefall, and over to a small weather station high on the glacier's east side. Crampons probably weren't necessary but I wore them anyway. Black clouds, pregnant with moisture, sailed north into the Alaska Range, banging into ridges, sticking to peaks. Tufts of tundra covered the lower reaches of the valley. Dall sheep—tiny white dots in the distance—clung to their high green ground. Everywhere else, the rock was bare, grey, primordial. Angular fragments of granodiorite, loosened by the rain, clinked down the surrounding cliffs. Snow began to collect a thousand feet above us. The Pleistocene was settling back around our shoulders like a shawl. It was August.

⋏ *Gulkana Glacier flowing down from the Alaska Range.*

◄ *Terminus of Gulkana Glacier, north of Paxson, Alaska Range.*

◄◄ *Snow collecting on the upper reaches of Nebesna Glacier on the flanks of Mount Blackburn, Wrangell-Saint Elias National Park and Preserve.*

⋏ *Dall Sheep in the Alaska Range, Denali National Park and Preserve.*

3

Dennis Trabant had recorded the glacier's outline with a GPS receiver as we walked its lower perimeter. Rod March was a mile away, already at the weather station, fussing with the anemometer and precipitation gauges. At that time, the US Geological Survey had exactly two glaciologists working full time in all of Alaska—Dennis and Rod. I was along for the ride, learning their glacial world from the ground up.

Crossing the glacier, I wove a zig-zag path through a field of minor crevasses. These cracks extended up and down the glacier, parallel to its spine. The breaks were only twenty or thirty feet deep, separated by comfortably wide catwalks of surface ice. I walked first left, then right, then left, through this rectilinear maze—ten steps sideways for every one forward. I could hear water rushing below as I jumped two or three feet over the narrower crevasses. I was getting used to this place.

▼ Bare flanks above Gulkana Glacier attest to its dramatic shrinkage over the past 150 years.

We spent two hours at the weather station, changing antifreeze, repairing wind vanes, tinkering with batteries. I kept looking south, where even darker clouds roiled as the storm grew more ominous. Dennis and Rod patiently ignored my not-so-subtle hints about the approaching rain. Running low on optimism, I assumed that it would wash us off this glacier—or worse, turn into snow and trap us up here for the winter. They smiled, shrugged, and kept working as rain began to fall in earnest.

The maintenance chores were done by 5 PM. We buttoned up the station and headed off the moraine. The truck was too many miles away to think about. Rain steadily filled my backpack as we marched down the glacier. The Nikon camera submerged in the front compartment blew bubbles of disbelief at my carelessness. My crampons made a satisfying s-c-r-i-t-c-h with each bite into the ice. Unfortunately for the camera, I scarcely noticed the rain, mesmerized instead by this wild white world. Glacier—this amalgam of an all-but-infinite number of snowflakes. Glacier—this irresistible force leveling the mountains even as I watched. Glacier—this great sleeping beast beneath my feet.

▼ *The terminus of Gulkana Glacier has been steadily retreating as the glacier loses volume. Dennis Trabant and Rod March use GPS technology to monitor the retreat.*

ICE. Three-quarters of all fresh water in the world is locked up as ice. Six million square miles of the Earth's surface lie shrouded beneath a silent white blanket. Near the end of the Pleistocene ice ages twenty thousand years ago, three times more ground was ice-covered—at least a quarter of the Earth's land surface. Glacial ice exerts a profound influence on the shape of the world in which we live—gouging the Great Lakes, dismantling the Himalayas, scooping out the fjords of Norway, beveling the American Midwest. Nowhere else in North America can we find better examples of these on-going processes than throughout the state of Alaska.

Gulkana is only one of Alaska's one hundred thousand glaciers. This number, at best a crude guess, rises and falls as the climate changes and icefields expand or contract. Ice covers 29,000 square miles of this state—about five percent of its total land surface. And where there is no ice, you're likely to find evidence of past ice ages—land forms that have been whittled down or heaped up by the great ice sheets that buried half of Alaska eighteen thousand years ago.

Why is there so much ice in Alaska? Why not North Dakota, where it's just about as cold? I learned the answer to this question the hard way as I flew, floated, and hiked through Alaska's glacier country. Storms were always brewing and my boots were always wet. If precipitation isn't falling yet, it will soon. Snow is the first, second, and third ingredient in any glacier pie.

A quick squint at a map shows glaciers rimming the coast of the Gulf of Alaska. The gulf is home to the semi-permanent Aleutian Low, a predictable area of low atmospheric pressure that generates so much of the weather that goes on to soak the rest of the United States and Canada. These storms, spinning counterclockwise, take their first landfall punches against the coastlines of southeastern Alaska and Prince William Sound. Rainfall at Yakutat averages 130 inches a year. The Juneau Icefield receives 100 feet of snow every winter.

◄ *Fog clings to the Takhinsha Mountains west of Haines.*

A L A S K A

Anchorage

ALASKA RANGE

Kenai

Peninsula

Harding
Icefield

Prince
William
Sound

Bering
Glacier

ST. ELI

CHUGACH RANGE

Alaska Peninsula

Kodiak Island

G U L F

The majority of Alaska's glaciers are close to the coast. All begin
in mountain ranges and flow down to lower altitudes,
sometimes reaching sea level. There are five major regions of
glaciers within the state. The Coast Mountains adjacent to the
Inside Passage are home to the Juneau and Stikine icefields. The
Fairweather Range traps precipitation that blows in off the
Pacific, and sheds its load of snow into Glacier Bay National
Park and Preserve. The St. Elias and Wrangell mountains spawn
the state's largest glaciers. The Chugach and Kenai mountains
surround Prince William Sound. The Alaska Range separates the
southern coastal regions from Alaska's drier interior. All of these
mountain ranges effectively strain precipitation from any
passing storm. Glaciers also stretch down the Aleutian Chain,
and others hide within the peaks of the northern Brooks Range.

P A C I F I C   O C E A N

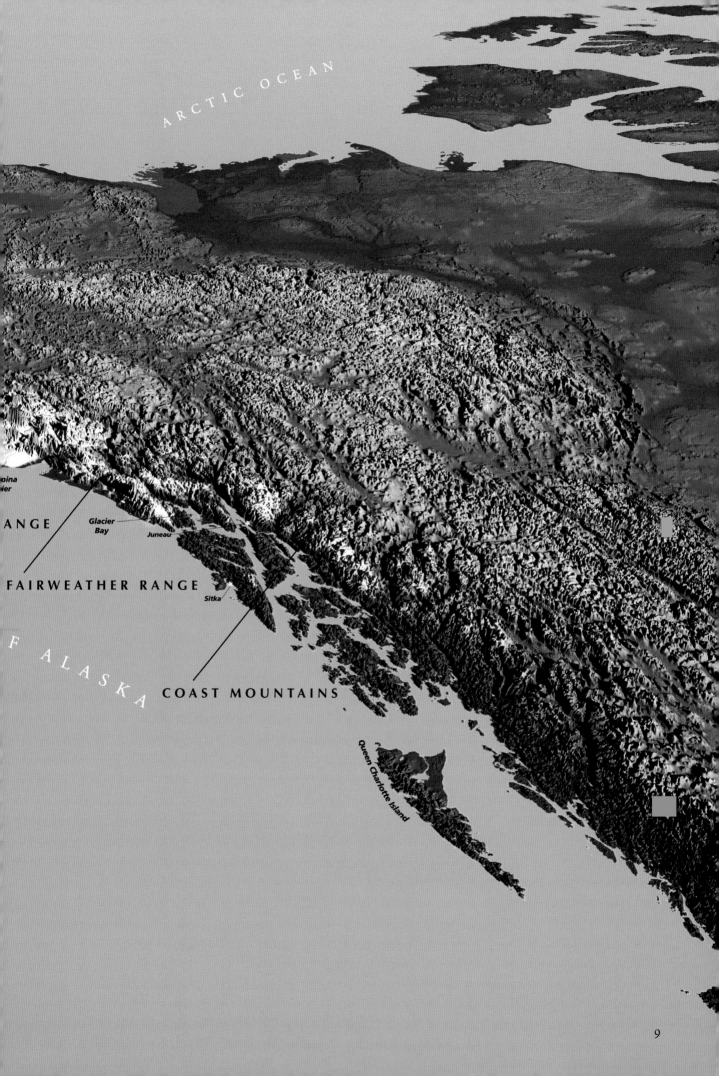

ARCTIC OCEAN

oina
ier

ANGE

Glacier
Bay

Juneau

FAIRWEATHER RANGE

Sitka

F ALASKA

COAST MOUNTAINS

Queen Charlotte Island

So what are glaciers? The answer is quite simple. Glaciers are thermodynamic systems that 1) absorb precipitation and rock debris; 2) get stirred around by gravity, solar energy, and geothermal heat; and 3) spit out water and rock. That's it: What goes in comes out. Perhaps I should start at the beginning. . . .

When snow falls, it doesn't always get around to melting right away. A snowfall in the desert Southwest may last until morning when the first sunlight strikes. Snow that falls high in Washington's Cascade Mountains accumulates throughout the winter, to melt in late spring. But there are glaciers in the Cascades, so not all of it melts. And in Alaska, much of the snow that falls doesn't melt from one year to the next. Instead, snowflakes pile on top of snowflakes in steadily thickening layers. Once buried, the layers are insulated and far less likely to melt. We don't have a glacier on our hands yet, just a block of ice. Glaciers are rivers of ice. Rivers, not lakes. A glacier first exists when a static accumulation of ice becomes thick enough to move laterally under the influence of gravity. Now we're getting somewhere, literally.

▲ Melting ice at the terminus of the Mendenhall Glacier decays into unstable caves, Tongass National Forest.

Despite the tremendous volume of ice that still covers parts of Alaska, seeing glaciers here can take a bit of effort and expense. You can count the state's major highways on one hand. Nevertheless, a few glaciers can be reached by road. If you've driven out of Anchorage along Turnagain Arm looking for Portage Glacier, you may well arrive at the Begich, Boggs Visitor Center and ask, "What glacier?" The view through the visitor center's window is of a beautiful aquamarine lake surrounded by soaring mountains. But no glacier. There really was a glacier here in the past—an icy ramp folks used to portage between Prince William Sound and Turnagain Arm before World War II. But in the 1980s ice began to spall off the glacier faster than it could be delivered. Portage Lake clogged with icebergs, and the glacier's face retreated to its current position three miles up-valley. A boat will take you there. Even better, drive through the tunnel to Whittier and hike the old portage trail that climbs 400 feet to a fine view of the lake and what's left of the glacier.

Continue past Portage toward Seward, and you'll have a chance to visit Exit Glacier, part of Kenai Fjords National Park. Exit is

◄ Bench Glacier, northeast of Valdez, spills north out of the Chugach Mountains, Chugach National Forest.

Take the Seward Highway south of Anchorage to visit Portage and Exit glaciers.

CHANNEL

Hope

Whittier

CHUGACH
NATIONAL
FOREST

PORTAGE
GLACIER

Sterling Highway

Soldotna

KENAI
NATIONAL
WILDLIFE
REFUGE

Seward Highway

KENAI
FJORDS
NATIONAL
PARK

Seward

EXIT GLACIER

KENAI
NATIONAL
WILDLIFE
REFUGE

one of thirty-two glaciers that spill down from the Harding Icefield. A short walk takes you to Exit's snout. Signs along the trail mark the glacier's recession since the nineteenth century. This terminus lies on a rumpled gravel plain dotted with fireweed, one of the first plants to invade newly exposed soil as glaciers retreat. What a remarkable landscape! Stand near (but not too close to) the glacier and look up: All that ice is slithering and sliding inexorably down toward you. A couple of hundred years ago, the location of the terminus was stationary, implying that ice was being delivered from above at exactly the same rate that it was melting below. But Alaska is a warmer place these days, and the melting has accelerated. The result is that the terminus is retreating up-valley.

A number of other glaciers can be seen from the road. Matanuska Glacier is two hours by car northeast of Anchorage. It winds twenty-seven spectacular miles down from its lofty source in the Chugach Mountains, but the roadside view

◄ *Tour boat cruises past Surprise Glacier in Prince William Sound, Chugach National Forest (above); Matanuska Glacier writhes down from the Chugach Mountains above the Glenn Highway (below).*

these days is one of as much gravel as glacier. Farther east near Valdez, Worthington Glacier would take out the Richardson Highway if it started advancing back into its old Ice Age position. And the Copper Highway doesn't go much of anywhere beyond Cordova, but it sure has a good time getting there. Along the way there are lovely views of the Sheridan and Childs glaciers.

Many of Alaska's most impressive glaciers end in the sea. Boats that berth in Whittier, Valdez, Seward, and Petersburg will take you to the tidewater termini of glaciers in Prince William Sound, Kenai Fjords, and Stephens Passage. Try not to be distracted by the whales and sea otters; remember, you're here to see glaciers. Cruise ships with populations of small cities regularly glide into Glacier Bay on their way up or down the Inside Passage. Alternatively, you can climb into a kayak and spend a week or two paddling through the bergy bits and brash ice on your own. Nothing, absolutely nothing, will beat the view of Surprise Glacier from that front-row kayak seat, as your boat gently rocks on waves from the last crashing iceberg.

Airplanes open up the brave new world of glaciers throughout Alaska. Fixed-wing tour flights rise like hatches of mayflies out of every airport within a hundred miles of any glacier you'd wish to see. Mendenhall Glacier looms above its US Forest Service Visitor Center just outside Juneau. Short trails lead down to the lake and waterfalls in front of the glacier; longer ones take you up the glacier's west side. But many people see the Mendenhall from helicopters that sometimes land up on the ice. Radial-engine DeHavilland Beavers on floats will taxi to your cruise ship in the Gastineau Channel, and then fly over Taku Glacier and the Juneau Icefield. And scramble for that left-side window seat on your flight home. From 35,000 feet, the icy view of Alaska's glaciers will stretch out for an hour and a half on a clear day.

➤ *The Fairweather Range above Glacier Bay National Park and Preserve is a vertical world of ice and rock.*

We sat in our kayaks before Lamplugh Glacier. The water was littered with ice of all sizes. Tom Bean, once a park ranger at Glacier Bay, regaled me with tales of crouching like this in front of Muir Glacier when a half-mile-long ice face peeled off. I back-paddled a bit as he told the story. We watched in silence for a while. Then a large sheaf of ice spalled, dropping in slow-motion toward the water. It crashed through the glassy surface and generated a steep breaking wave that rushed toward us.

Twenty or thirty yards away, the wave smoothed into a great undulating swell. Our rectilinear world was suddenly transformed into something very non-Euclidean: What had been flat suddenly became sinusoidal. Ten-foot blocks of ice sloshed above and then below us as the kayaks pitched up and down on the wave. I felt at once foolish and delighted to be there. The water calmed and we lingered in front of Lamplugh, gleaming and golden in the last rays of sunset.

Harding Icefield.

P

Exit Glacier Ro

Interpretive
Shelter

Harding
Icefield Trailhead

Nature Trail

End of Trail
Mile 3.86

EXIT GLACIER

100

1500

2000

2500

4000

3000

2000

1500

1500

2000

2500

5

Exit is one of many glaciers that pour down from the Harding Icefield.

Kenai Fjords National Park

Exit Glacier is a landlocked glacier streaming out of the Harding Icefield within walking distance from the visitor center and the end of the road. Over the last 50 years, Exit Glacier has been retreating up the valley, about 1,800 feet since 1950. Exit Glacier is the only part of Kenai Fjords National Park that is accessible by road; about 100,000 people visit this glacier annually.

Best Viewing Access

Walk to Exit Glacier on your own or participate in a ranger-led walk offered twice daily during summer months.

What Else to Do

See the Harding Icefield; a challenging 7-mile round-trip hiking trail runs approximately parallel to Exit Glacier.

Take a boat tour of Kenai Fjords for views of calving tidewater glaciers, not to mention the otters, whales, and seabirds abundant in the fjords' rich marine environment.

Visit the Kenai Fjords National Park Visitor Center in Seward or the Nature Center at Exit Glacier to learn more.

Fireweed is one of the first plants to colonize the rocky soil that is exposed as Exit Glacier retreats.

Getting There

Exit Glacier is just 12 miles from Seward. The Herman Leirer Exit Glacier Road leaves the Seward Highway just north of town and takes visitors 8.5 miles to the Exit Glacier Nature Center; an about three-quarters of a mile walk loops past the snout of the glacier, giving excellent views. This is the only road-accessible entrance to Kenai Fjords National Park.

Website

www.nps.gov/kefj

Seward Highway

Paradise Creek

Chapter Two

WILDNESS UNSPEAKABLY
PURE AND SUBLIME

On July 15, 1741, Georg Steller paced the deck of the *St. Peter*, a brig-rigged two-master captained for the Russians by Vitus Bering. After six weeks at sea, the overcast swirled open just long enough for him to shout, "LAND!" The following day, the clouds again parted to reveal a glistening white peak of immense grandeur, which the explorers named after the saint-du-jour: Saint Elias. Leif Erickson and Christopher Columbus had discovered the New World by sailing west. Bering was the first European to reach this land by sailing east.

The crew of the *St. Peter* initially sighted Alaska near Yakutat Bay. They sailed up the coast, passing two glaciers now named Malaspina and Bering. Five days after first glimpsing the coast, they made landfall on Kayak Island to fill the ship's water casks. Steller, the trip's naturalist, recognized a crested blue bird that he recalled seeing in the recently published Catesby's *Natural History of the Carolinas and Florida*. He leapt instantly and accurately to the conclusion that this indeed was the New World. (We know that bird today as the Steller's jay, *Cyanocitta stelleri*.)

The Russian crew eventually returned home with burdensome baggage: news of unlimited numbers of animals cloaked in the most wonderful fur. Almost a half-century passed before Europeans would come this way for anything more enlightened than plundering sea otter pelts. Alesandro Malaspina outfitted two Spanish corvettes for the purpose of mapping the coast, hoping to find the elusive (indeed, non-existent) Northwest Passage through to the Atlantic. In 1791, he wove through icebergs into the back of Yakutat Bay. He pushed as far east as he could, but was finally thwarted by Hubbard Glacier's wall of tidewater ice. He left, naming this inlet *Desengaño*, now loosely translated as Disenchantment Bay. The great mass of ice just north of his course now bears his name, Malaspina Glacier.

▲ *Wrangell Mountains east of Parka Peak, Wrangell-Saint Elias National Park and Preserve.*

◄ *Contorted moraines at the lower end of Malaspina Glacier, Wrangell-Saint Elias National Park and Preserve.*

◄◄

Bering Glacier empties into the Gulf of Alaska, Bureau of Land Management.

George Vancouver sailed this way in 1794, mapping the coast and Inside Passage for the English. He concentrated on dodging icebergs as he passed the mouth of Glacier Bay. Instead of open water, he found a wall of ice 150 feet high along the north shore of Icy Straits. Vancouver's maps offer no hint of Glacier Bay, which now extends 65 miles beyond the ice cliff he saw—the entire bay was filled with thousands of feet of glacial ice then.

The Hoonah Tlingit tell of a young woman, Shaawatsèek´, who called a glacier down upon her village inside Glacier Bay hundreds of years ago, and then of the old woman Kaasteen, who offered to stay in the glacier's path by way of supplication. Perhaps Shaawatsèek's impetuousness caused ice to advance at the beginning of the Little Ice Age, a world-wide, three-century-long cold spell. And who knows—without Kaasteen's good intentions, the Tlingit might not have been able to return after the ice's retreat from Glacier Bay around 1850 at the end of the Little Ice Age.

The newborn science of glaciology stood on wobbly legs in the mid-nineteenth century. In his classic 1833 text, *Principles of Geology*, Charles Lyell toyed with the notion that some peculiar rocks in England had been carried great distances by glaciers. A year later, Jean de Charpentier presented evidence of Alpine glaciation to the Swiss Society of Natural Sciences. Entrenched in their belief that all landscapes were the product of the Biblical flood, the society's members dismissed these new notions. Louis Agassiz studied piles of glacial rubble (moraines) in the Rhone Valley in 1836. He was convinced that the Earth had previously experienced an "epoch of intense cold" when glaciers swept down over much of the Northern Hemisphere sometime in the past. The work of these men was slowly accepted by scientists on both sides of the Atlantic.

➤ *As it reaches Disenchantment Bay, the surface of Hubbard Glacier becomes deeply fractured.*

John Muir was already convinced that ancient glaciers had carved the glories of Yosemite Valley by the time he came north in 1879, seeking modern examples of real-world ice. (The United States had purchased Alaska from the Russians for two cents an acre only twelve years before his arrival.) In a way, though, it was he who retreated into the past instead. Bobbing in an open canoe amidst icebergs, he might as well have been visiting the Pleistocene Ice Age. His Tlingit guides shook their heads as Muir relentlessly pushed on, paddling into the teeth of Glacier Bay's October winds and rain. Weather notwithstanding, Muir jubilantly wrote about "the crystal bluffs of two vast glaciers, and the intensely white, far spreading fields of ice, and the ineffably chaste and spiritual heights of the Fairweather Range, now partly revealed, the whole making a picture of icy wildness unspeakably pure and sublime."

Muir's popular writing opened Glacier Bay and southeastern Alaska to a wave of exploration and tourism. Henry Fielding Reid carefully mapped Glacier Bay in 1890 and 1892. By then, the runaway retreat of the bay's glaciers had exposed miles of virgin territory. The air shook with the rumbling thunder of glacial calving and the water was lousy with floating ice—no small threat to Reid's rowboat. He described the "awful roaring, tons of water streaming like hair

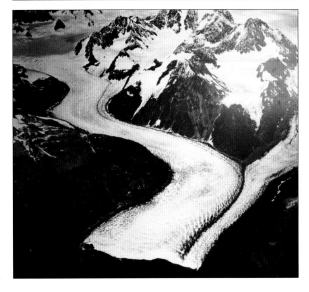

▲ *Historic glacier photos from National Snow and Ice Data Center archives help scientists see changes. From top, Muir Glacier, Falling Glacier, and Meares Glacier.*

25

down the sides, while they heave and plunge again and again before they settle in poise and sail away as blue-crystal islands."

John Muir returned to Alaska in 1899 aboard the *George W. Elder*, a steamer chartered by railroad magnate Edward Harriman. What started as a rich man's lark turned out to be a singularly successful scientific expedition. Muir was in the company of some of the most prominent scientists, writers, and artists of his day—John Burroughs, William Dall, G.K. Gilbert, George Grinnell, C. Hart Merriam, Edward Curtis, and Frederick Dellenbaugh to mention a few. The trip explored Glacier and Disenchantment bays, Malaspina and Hubbard glaciers, eventually reaching the Bering Straits. The *Elder* made its way into the northwestern-most corners of Prince William Sound. There, top-heavy with Eastern academicians, the group bestowed such names as Harvard, Yale, Bryn Mawr, Amherst, Dartmouth, Vassar, Wellesley, and Holyoke upon glaciers within College Fjord. They pushed through swirling tidal currents at the foot of Barry Glacier, where Harriman Fjord was named in the expedition's wake.

Eliot Blackwelder, Ralph Tarr, and Lawrence Martin were scientists who mapped glaciers from Yakutat Bay to Prince William Sound at the beginning of the twentieth century. Blackwelder spent the summer of 1907 between Yakutat Bay and the Alsek River. Tarr and Martin were particularly interested in the glacial effects of an intense earthquake that struck this coastline just after the Harriman expedition departed in 1899. These men explored a huge country on foot and horseback, rarely bothering in their publications to mention the arduous day-to-day details of raw survival in a wild landscape. William Field brought his camera to Alaska in 1926, reoccupying locations from which Henry Reid and the Harriman party had photographed glaciers decades earlier. Field, along with Maynard Miller, started the Juneau Icefield Research Program (JIRP) after World War II. The JIRP continues glaciologic research and training today on the Mendenhall, Taku, and surrounding glaciers above Juneau.

▲ *Hubbard Glacier (above left) and Turner Glacier (below right)*
pinch the upper end of Disenchantment Bay. Russell Fjord (above right)
is almost completely blocked by Hubbard Glacier.

▲ Johns Hopkins Glacier, 1964, photographed by Austin S. Post.

The US Geological Survey has a long history of glacial research in Alaska. One of the Survey's first and brightest lights was Austin Post. A high-school dropout, Post had gathered enough casual experience with glaciers and aerial photography to submit a research proposal to the National Science Foundation in 1960. He listed himself as the project's Senior Meteorologist, managing to misspell both "senior" and "meteorologist." Not easily put off, the NSF funded Post to fly over and photograph as many of Alaska's glaciers as he could catch in his viewfinder. The photographs were stunning, revealing much about the structure and behavior of glaciers never before recorded. By the time the Good Friday earthquake rocked Anchorage in 1964, Post's aerial photography was well recognized as an indispensable tool for recording glacial and geologic events. He stayed on with the USGS, amassing a library in excess of one hundred thousand photographs.

Post understood that airplanes can tremendously speed the reconnaissance of glaciers in Alaska's forbidding terrain. But others knew that many icy particulars

can't be studied from the air. Sometimes you just have to strap on the crampons and start walking. In the early 1960s, Troy Péwé drove as close as he could to Gulkana Glacier and then slogged the remaining miles to reach the ice. Larry Mayo hired pilots who could fly him into Wolverine Glacier on the Kenai Peninsula, where he would spend weeks in its wilderness. These men carefully recorded what was actually happening down on the ice—weather conditions, snow accumulation, and stream runoff. They repeatedly surveyed surface elevation and ice velocity from year to year. Dennis Trabant and Rod March continued these studies of Gulkana and Wolverine glaciers into the twenty-first century for the USGS. Long-term uninterrupted records are invaluable tools in the ongoing study of glaciers and the climates that they reflect.

▲ *Tyeen Glacier, 1964; Douglas Glacier, 1960, both by Austin S. Post.*

I got fairly good with a shovel on Wolverine

Glacier. Dennis and Rod called down encouragement as I flung yet another load of snow out of the ten-foot-deep pit—they seemed to like this Tom Sawyer approach to glaciology. The real science began when the digging stopped. Rod sampled snow at intervals along the wall, measuring density and water concentrations. Dennis used a metal detector to locate last year's stakes, buried deep beneath the winter's snowfall. I was surprised to find that ice temperatures hovered right at the freezing point, no matter how deeply we dug or drilled into the glacier. At night, we crowded into the twelve-by-sixteen-foot hut that Larry Mayo had built three decades earlier.

After dinner, sitting on the doorstep, I watched the red embers of sunset fade into a purple glow that bathed the neighboring Sargent Icefield. As darkness fell, I marveled at the intense silence that enfolded this wonderful world. Only then did I begin to fathom why scientists had been returning to study these glaciers for so many years.

◄ *Dennis Trabant and Rod March taking measurements in a snow pit on Wolverine Glacier. Far left, Wolverine Glacier on the Kenai Peninsula.*

Glacier Bay National Park and Preserve

There are glaciers galore in this dramatic landscape of water, rock, and ice, as well as opportunities to see first-hand the creation of new life on scoured rock. The rich marine environment hosts humpback whales during summer months.

Best Viewing Access

Glacier Bay is best seen by boat. Daily boat tours operate in the summer months for those folks not arriving on a cruise ship or without their own boat. From park headquarters at Bartlett Cove, the boat journey is an all-day adventure, as the tidewater glaciers are up-bay about 60 miles. National Park Service Rangers accompany cruise ships and tour boat trips.

What Else to Do

Kayak or flight-see Glacier Bay; trips can be arranged through local providers.

Watch for whales; breeding humpbacks come to Glacier Bay to feast in the summer.

Getting There

A short air flight from Juneau to Gustavus brings most independent travelers to Glacier Bay, and most cruise lines plying the waters of southeastern Alaska stop in Glacier Bay. Glacier Bay is not accessible by road or the ferry system.

Website

http://www.nps.gov/glba

GRAND PACIFIC GLACIER

MARGERIE GLACIER

LAMPLUGH GLACIER

REID GLACIER

MUIR INLET

CASE

GLACIER BAY

BRADY ICEFIELD

BRADY GLACIER

Chapter Three
A GLACIER
FOR EVERY PURPOSE

A GLACIER IS A GLACIER, RIGHT? Just a big block of ice. Wrong. You got your warm glaciers and your cold glaciers. You got your glaciers in mountains and your glaciers in valleys. Glaciers found near the ocean are different than those far inland. There are good reasons to make these distinctions. Each type of glacier forms, behaves, and melts in its own rhythm.

At first glance, it's natural to think that all glaciers are cold. Some *are* "cold," a term that implies a number of interrelated aspects of ice, including temperature regime, geographic setting, and mechanical properties. Cold polar ice in Antarctica and Greenland exists where atmospheric temperatures are almost always far below the freezing point of water. Free water is virtually nonexistent on these ice sheets. For the most part, polar glaciers shed ice not by melting, but by moving great distances over land and then launching tabular icebergs along their margins into the sea. Cold glaciers tend to form far from the warming influence of oceans. They are likely to be *continental*, rather than *maritime*, with less moisture available for precipitation.

The interior of Antarctica is among the driest places on earth, typically receiving two inches of snow a year. Precipitation rates are the bottleneck in polar ice development—the long-term health of a continental glacier is most directly related to fluctuations in precipitation rather than rates of summertime melting. Cold glaciers move very slowly because they accumulate so little snowfall, and because their bases are frozen to the underlying bedrock. Meserve Glacier in Antarctica crawls along at one-quarter inch per year, almost slow enough to challenge its description as a glacier rather than stagnant ice.

"Warm" glaciers are altogether different. They exist in temperate environments like coastal Alaska, where precipitation is plentiful. Warm glaciers are not frozen to their bed. They slide downhill rather than sticking. Warm glaciers in Alaska typically gallop along at rates approaching 1,000 feet per year, orders of magnitude faster than their cold Antarctic cousins. Depending on the season, a warm glacier will have liquid water running across its surface, along its bed, and in channels through its interior. Of course, *warm* is a relative term.

⋏ An alder leaf hitches a ride on Root Glacier, Wrangell-Saint Elias National Park and Preserve.

◄ Hanging glaciers above Backside Glacier southeast of Mount Hunter, Denali National Park and Preserve.

◄◄ The Hole in the Wall Glacier flows out of the coastal mountains of the Tongass National Forest in southeastern Alaska, splintering away from Taku Glacier just before reaching sea level.

▲ Rock glacier near Nelson Mountain, east of Chitina.

Some heat enters a warm glacier from geothermal sources at its bed, but larger amounts enter from above, contributed by direct solar radiation during the summer melting season. When rainfall manages to flow into a glacier, it can be trapped inside and freeze, effectively warming the surrounding ice. Warm glaciers generate friction when they slide over their bases, introducing yet more heat. The end result of all these thermal inputs is that most of Alaska's glaciers—certainly all of its maritime glaciers—hover surprisingly close to their melting points. Think of it: thousands of cubic miles of ice, all within a few tenths of a degree of melting.

Polar ice sheets are vastly older than Alaskan valley glaciers. The continental "slow" ice in Antarctica is truly ancient—five hundred thousand years if it's a day. But maritime snow accumulates and flows much faster. The age of most ice in Alaska is measured in hundreds rather than thousands of years.

Antarctica and Greenland have *ice sheets* and *ice caps*—expanses of ice so thick and wide that their motion is not constrained by underlying topography. Ice sheets spread outward, expanding radially in all directions. Alaska isn't submerged beneath these gigantic amoeba-like sheets of ice. Instead, glaciers here are of such a size that their motion is everywhere influenced by the mountains and valleys in which they exist. The largest glacial features in Alaska are *icefields*. The Bagley Icefield, shared by Alaska, British Columbia, and the Yukon, is an interconnected phalanx of glaciers in the vicinity of Mount St. Elias. Over 100 miles long, it acts as the primary source for a multitude of smaller glaciers that flow downhill in all directions. The Bagley Icefield is constrained by ridges that spin off the St. Elias Mountains, forcing the ice into an east–west orientation.

A *cirque glacier* is a small body of ice clinging to a niche high on a mountainside. The cirque is a hollowed-out bedrock bowl that acts as the glacier's accumulation zone. Snow can, of course, fall directly onto a cirque glacier, but more often than not, it arrives indirectly as a veneer of windblown ice crystals. Ice might flow out of the bowl but not reach the valley floor below, in which case it is called a *hanging glacier*.

Rock glaciers are a thick stew that contains as much rock as ice. These odd glaciers are found in mountains that actively shed a great deal of loose rock onto talus slopes. Water percolates into the talus, and rocks that were initially in direct contact with each other begin to "float" as the water freezes and expands. On the surface, a rock glacier looks a lot like your everyday pile of rocks. But the action is all on the inside. As rocks and ice pile up, their cumulative weight eventually exceeds the shear strength of the interstitial ice. Then the whole glacier begins to deform plastically and slowly flow downhill.

A *piedmont glacier* is a thick continuous mass of ice resting on relatively flat land near the base of a mountain range. Piedmont glaciers are fed by other glaciers that deliver ice from nearby mountains. The Bagley Icefield spawns two piedmont glaciers, Bering and Malaspina. Each is the size of one of those poor Eastern Seaboard states that are always picked on for being so small (Rhode Island in this case, if you must know). Both of these behemoths have histories of erratic behavior, called surging, which is recorded in taffy-like swirls on their surfaces.

▲ *Redoubt Volcano, Lake Clark National Park and Preserve.*

A *valley glacier* is a python of ice gliding down a topographic cleft. This type of glacier may sport a single trunk or, more often, consolidate subsidiary glaciers into a final great trunk. Some valley glaciers originate in cirques; others are fed by icefields at their higher elevations. Ice thickness, measured midway across the glacier, will usually be about one-half its width. As they grind themselves into progressively deeper gorges, valley glaciers are particularly sensitive to local bedrock geology. If a zone of weakness exists—such as a major fault—the glacier will preferentially align itself with that underlying geological fabric.

Most of Alaska's frequently visited glaciers are of the valley persuasion— Mendenhall, Matanuska, Worthington. Kennicott is a classic valley glacier. It streams down from the southeast face of Mount Blackburn in the heart of Wrangell–St. Elias National Park and Preserve near McCarthy. The drive in from Chitina is pleasant enough; the 60 teeth-rattling dirt miles have actually been somewhat improved lately. McCarthy's charm lies largely in its cultivated inaccessibility: you can't get there from here, at least not by car, because no automobile bridge yet spans the Kennicott River. You *can* walk the last mile into McCarthy. The Kennicott mill and townsite are four more miles up the valley. A little farther, you can stroll across the Root Glacier to its confluence with the

Schematic of valley glacier.

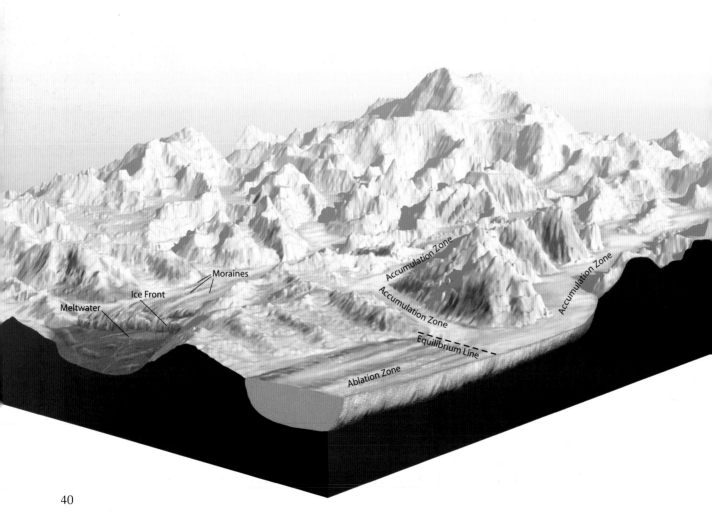

Moraines

Ice Front

Meltwater

Accumulation Zone

Accumulation Zone

Accumulation Zone

Equilibrium Line

Ablation Zone

Kennicott. Pick a good dry rock, take off your crampons, and sit a spell. Make sure that the black bear you waved to back in the alders hasn't followed you out onto the ice. Now look around. . . .

This valley is nothing short of spectacular. Bonanza Ridge sweeps down upon Porphyry Mountain, just this side of Sourdough Peak. Across the valley, Fourth of July Creek drains the north side of Fireweed Mountain. To the north, Donoho Peak hides Gates Glacier. Just to the right, Root Glacier tumbles over the impressive Stairway Icefall as it rushes away from Regal Mountain. The Kennicott Glacier weaves through this symphony of a landscape, orchestrating its

▲ *Sinuous medial moraines mark the path of Kennicott Glacier below Mount Blackburn, Wrangell-Saint Elias National Park and Preserve.*

41

mood, shaping its swales, gradually tearing it down, one rock at a time. The ice is everywhere impregnated with rock—the terminus is buried beneath vast piles of boulders and gravel. Long multicolored ribbons of morainal rubble wave like banners as they pass Packsaddle Island. The sound of running water is pervasive—noisy creeks play hide-and-seek along the glacier's margin. Surface water falls ominously into *moulins*, manholes on the surface of a glacier, that emit the deeply disconcerting noises of subterranean machinery. Rocks clatter into ravines as ice melts out from under them. This landscape is alive. It is pushy. Glaciers like the Kennicott are electrified with a wonderful sense of geologic urgency.

One more definition: *tidewater glacier*. When a glacier calves ice directly into seawater, it's known as a tidewater glacier. Calving, heralded by thunderclaps, can be awesome. Ice pillars

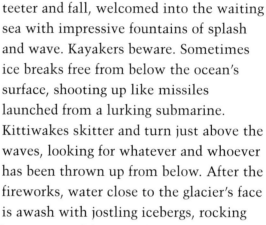

teeter and fall, welcomed into the waiting sea with impressive fountains of splash and wave. Kayakers beware. Sometimes ice breaks free from below the ocean's surface, shooting up like missiles launched from a lurking submarine. Kittiwakes skitter and turn just above the waves, looking for whatever and whoever has been thrown up from below. After the fireworks, water close to the glacier's face is awash with jostling icebergs, rocking toward a temporary buoyant equilibrium. The bergs melt quickly, rolling this way and that for a few days before being swallowed by the sea.

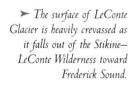

▲ *Icebergs from Columbia Glacier, in the Chugach National Forest, make their way into Prince William Sound.*

➤ *The surface of LeConte Glacier is heavily crevassed as it falls out of the Stikine– LeConte Wilderness toward Frederick Sound.*

Unlike ice in Antarctica, Alaskan tidewater glaciers do not push floating ice shelves out over ocean water. These glaciers are warm and lack the internal integrity necessary to form stable tabular ice platforms stretching far out to sea. Instead, Alaskan glaciers extend seaward only when they are able to bulldoze a protective berm of rock and gravel in front of themselves. The glacier may thicken, eventually scouring out a floor hundreds of feet below sea level, but only if its calving face is relatively protected by that seaward berm. Indeed, if some element of glacial equilibrium changes for the worse—less ice or faster melting—a tidewater glacier will catastrophically retreat from the sea.

It's no secret that most of Alaska's glaciers have suffered terrible losses lately—more about that and climate change later. But a few tidewater glaciers are happily growing seaward today, oblivious to the headlong retreat of most others. Hubbard Glacier has a

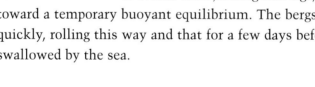

tidewater face that rises 300 feet into the air and stretches six
miles across Disenchantment Bay. Twice in twenty years, Hubbard
has pushed far enough into the bay to cut Russell Fjord off from
the sea, creating a 50-mile-long lake with a surface 60 to 80 feet
above sea level. And twice the temporary ice dam has broken with
floods of far greater discharge than the Mississippi River, rushing
through a chute a mere 600 feet wide. The bottoms of
Disenchantment and Yakutat bays are littered with evidence that
Hubbard Glacier has advanced far beyond its current position in
the last few hundred years. It is likely to do so again.

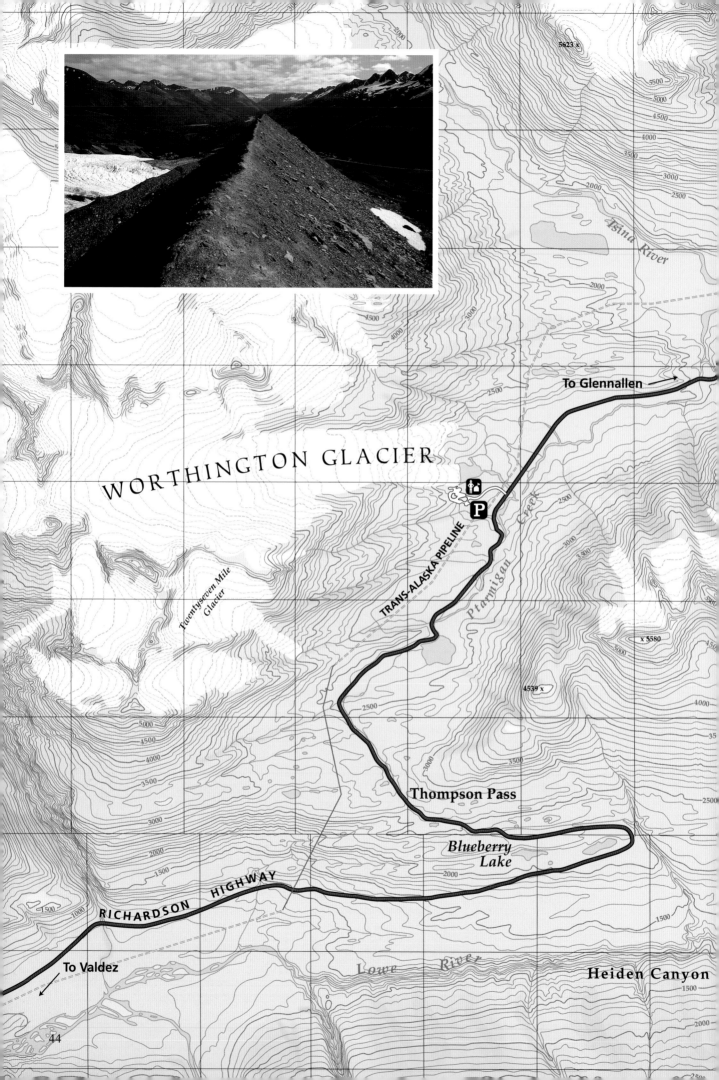

WORTHINGTON GLACIER

Twentyseven Mile Glacier

Isina River

To Glennallen →

TRANS-ALASKA PIPELINE

Ptarmigan Creek

5623 x

x 5580

4539 x

Thompson Pass

Blueberry Lake

RICHARDSON HIGHWAY

Lowe River

Heiden Canyon

To Valdez →

Worthington Glacier
State Recreational Site

Worthington Glacier is a nice example of a retreating valley glacier, easily seen by any travelers along the Richardson Highway between Valdez and Glennallen. The glacier's terminal moraine lies just off the road with excellent interpretive panels and viewing opportunities for people of all ages and abilities.

Tundra plants

Best Viewing Access

An interpretive trail with exhibits leads around the base of the glacier from the parking lot on a partially accessible trail.

What Else to Do

Follow the 1-mile-long Moraine Trail for a closer view of Worthington. This is a steep and, in places, narrow path; only the sure-footed and well-booted should attempt this option. Those who do will be well-rewarded with excellent views of blue ice and the surrounding alpine landscape.

Getting There

Mile 28.7 on the Richardson Highway between Valdez and Glennallen. Educational bookstore, interpretive exhibits, bathrooms, and an off-the-highway parking lot.

Website

http://www.alaskanha.org/worthington-glacier-state-recreation-site-valdez.htm

◄ *The trail up Worthington Glacier follows the knife-edge of a lateral moraine.*

▼ *The mid-reaches of Worthington Glacier are fairly smooth. Crevasses on the right begin to form as the lower glacier falls toward its terminus above Ptarmigan Creek.*

WORTHINGTON GLACIER

Chapter Four

THIS SCULPTED LAND

GLACIERS ARE IN THE BUSINESS of moving sediment. They grind up and spit out rocks of every size. *Glacial flour* is the fine powdery dust that imparts a milky appearance to water flowing below glaciers. *Erratics* are house-sized boulders hauled bodily away from their birthplace, then left to stand oddly alone—granite aliens in a limestone terrain. From flour to boulders and every size in between, glaciers manage to move massive amounts of rock. You may not be able to bring Malaspina to the mountain, but someday Malaspina will have brought every speck of Mount St. Elias to you.

Margerie Glacier picks away at the friable Fairweather Range and hauls 1.25 million cubic yards of rock down to the west arm of Glacier Bay every year. Muir Glacier, flowing south from the Takhinsha Mountains, delivers a third more to the bay's eastern arm. Some of this sediment is dumped at the calving face of the glaciers; more is rafted out into the deeper waters of their respective fjords. Johns Hopkins Inlet is just around the corner from Margerie; the bottom of this fjord accumulates 26 million cubic yards of sediment annually, a combination of ice- and stream-borne deposits. Bernard Hallet, a sedimentologist and glaciologist at the University of Washington, figures that Margerie Glacier is lowering its basin at the staggering rate of almost 2.5 inches per year. Geology doesn't get any better—or faster—than this.

▲ *Glaciers carried this block of granite down from the Alaska Range. The ice retreated, leaving the glacial erratic above McKinley Village.*

◄ *Small bergs fall from the face of Margerie Glacier. With each calving event, more sediment is dumped into Tarr Inlet at Glacier Bay National Park and Preserve.*

▼ *Johns Hopkins Glacier flows out of the Fairweather Range toward the west arm of Glacier Bay. Its surface is darkened by a heavy load of sediment eroded and transported from the surrounding mountains.*

◄◄ *Medial moraines on Carroll Glacier, Glacier Bay National Park and Preserve.*

Harriman Fjord was placid; our greatest fear had been that the kayak might sink if we rammed a sleeping sea otter. In other words, no fear at all. We paddled over a shallow outer moraine and beached on a postage-stamp island before Serpentine Glacier. I found twelve raspberries on the entire island. After setting up camp—tents, two stools, and a rainfly—we wandered over to the glacier. The tide was out. We secured the kayak and slogged around mudflats and sandbars. The terminus of Serpentine Glacier was lousy with rock. Water hissed through crevices and splashed down runnels. Gravel everywhere speckled the ice. Cobbles clinked in a soft continuous rain across the unstable slopes. A muddy stream spilled out of a

cave in the glacier's face, carrying a parade of bobbing cakes of floating ice.
Nothing grew here; we might as well have been on the moon.

The following day we sailed through a snowstorm of scissor-tailed
terns at the entrance to Barry Arm. The fjord was a cul-de-sac, ringed by
Coxe, Barry, and Cascade glaciers. We parked on a beach behind a rocky
barrier, knowing that waves washed over it once in a great while. The
graywacke—metamorphosed sedimentary rock—was deeply grooved,
sculpted by ice into the bulbous shape known as a roche moutonnée, with its
gentle slope facing upstream and steeper backside pointing down-glacier.
Falling pieces of ice cracked like rifleshots, breaking the silence with
unsettling regularity. Calving waves rippled across the fjord, constantly sieving
the beach's black sand. We spent hours in the sun, just wriggling our toes.
Bergy bits littered the beach, loitering like chilly chaperones waiting to melt.

◄ *The tides stranded a small block of ice on the
black sand beach across from Barry and Cascade
glaciers in Prince William Sound (above); Cascade
Glacier falls into the Barry Arm of Prince William
Sound (below), Chugach National Forest.*

▼ *A river rushes from the base of Serpentine
Glacier, spitting out large ice blocks that have been
gouged from the glacier, Chugach National Forest.*

W

HAT GOOD ARE GLACIERS ANYWAY? Maybe you operate a
sand and gravel company in Indiana. Ancient glaciers provided your bread and
butter. Maybe you're a wheat farmer in South Dakota. Glaciers first plowed your
fertile fields. Maybe you live in Los Angeles or New York or Liverpool or Hong
Kong or Buenos Aires. Modern glaciers have locked up enough ocean water to keep
you from blowing bubbles, 300 feet below what would be sea level. Maybe you're
the Loch Ness Monster. Glaciers gouged out the fjord that is your Scottish home.

Ice has shaped the world in which we live. Evidence for the earliest ice ages is fuzzy
but decipherable. Rocks called *tillites*, *mixtites*, and *diamictites* are poorly sorted
conglomerates that must have been deposited by the churning and grinding of
glaciers. Canada's Lake Huron region sports a 40,000-foot-thick sequence of such
sediments that date to the early Proterozoic, about two billion years ago—almost
half the Earth's age. During the Ordovician Period 400 million years ago, Africa lay
at the Earth's south pole, shivering beneath a 7,700,000-square-
mile ice sheet, considerably larger than the current expanse of
Antarctic ice. One hundred million years later, the irresistible
forces of plate tectonics bunched most of the Earth's continents
into Gondwanaland. An even larger ice sheet was then centered
on what is now southern Africa and Antarctica, with ice flowing
radially out in all directions.

The ice sheet that now covers Antarctica probably first developed
14 million years ago, after that continent separated from South
America. The opening of Drake's Passage allowed the Antarctic
Circumpolar Current to isolate Antarctica's climate from the
warming effects of the rest of the world. Mid-latitude glaciers
developed seven million years later with the rise of Asia's Tibetan
Plateau. As early as five million years ago, calving glaciers
contributed ice-rafted sediments to the 16,000 feet of tillites and
siltstones known as the Yakataga Formation off the coast of
southeastern Alaska. During the height of the Pleistocene, the
Laurentide Ice Sheet covered half of Alaska, all of Canada, and a
healthy portion of the northern United States—6,000,000 square
miles of North America.

When we speak of the Pleistocene ice ages, most people understand
that we're referring to the present glacial epoch that commenced
roughly 2.4 million years ago. Yes, *present*. The ice ages have been a
series of cold snaps separated by brief warm interludes. *Homo
sapiens'* entire experiment with agriculture and civilization has
taken place entirely within this latest interglacial respite.

After each glacial stand receded, it left a wide array of geomor-
phic features in its wake. Terminal and lateral moraines lie

abandoned in place, like children's toys left out overnight. Mendenhall Lake, so beautiful on a summer's evening, is impounded behind a terminal moraine that marks the glacier's extended position as recently as the 1930s. The high trail above Worthington Glacier follows the twisted spine of a lateral moraine. In places, the trail is little more than a razor's-edge atop the steep moraine. This portion of Worthington has thinned, and its surface is now well below the elevation at which this lateral moraine was deposited.

The classic U-shape is a clue that glaciers, not streams, carved a valley. Fingers of rock called *horns* and *arêtes* point skyward, left standing after glaciers have erased the lower surrounding topography. *Drumlins* are streamlined hills of glacial till that look for all the world like a pod of whales coursing just below the soil. *Kettles* are depressions left after stranded blocks of ice melt, lending a pock-marked appearance to a retreating glacier's outwash plain. Occasionally, a sinuous line of gravel called an *esker* is all that's left to indicate the former

▼ *Susitna Glacier, following the Denali Fault in the Alaska Range, gathers sidestreams of ice from the flanks of Mount Hayes. The glacier's combined moraines are contorted by the surging of first one sidestream and then another. During a surge, a glacier's velocity can suddenly increase by an order of magnitude or more.*

course of a sub-glacial channel. Both kettles and eskers can be seen along the Denali Highway as it crosses the outwash plain of Susitna Glacier, east of Cantwell.

Geology affirms that what goes around comes around. Sand is swept into piles that become sandstone that is buried to become quartzite. Mountains rise up and mountains are worn back down into sand grains. Glaciers certainly play a part in this revolving drama. Mountains help make glaciers, and glaciers help tear the mountains down.

➤ *Mount Fair-weather, Glacier Bay National Park and Preserve.*

Alaska's maritime glaciers hug the coast in a crescent from Wrangell around to the Aleutian Islands. Geologists have shown that the process of plate tectonics is rotating the floor of the Pacific Ocean counterclockwise relative to North America. The Pacific Plate's edge scratches northwest past Petersburg and Yakutat. The plate then dives back into the Earth in a trench southeast of the Aleutian Islands. This grinding is responsible for the Coastal, St. Elias, Wrangell, and Chugach mountains. The Chugach raise their 12- and 13,000-foot spine across the north end of Prince William Sound. A second tier of mountains—the Alaska Range—marks the path of the Denali Fault, an inland splay of the plate boundary. All of these mountains wring tremendous amounts of moisture from every passing storm—moisture that falls as snow and quickly builds up as glaciers.

Tectonic map of Alaska.

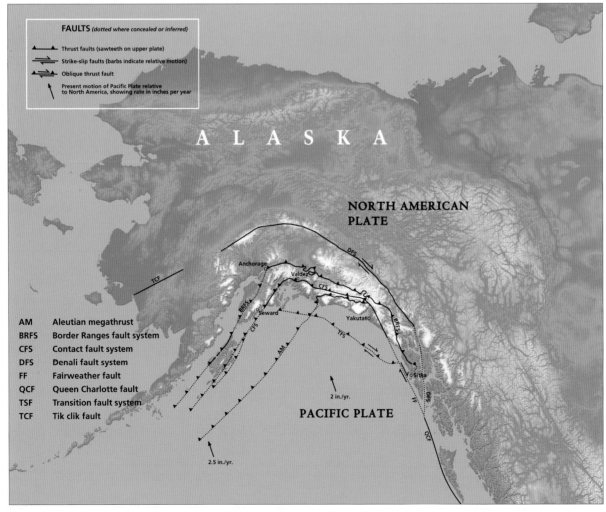

FAULTS (dotted where concealed or inferred)

▲▲▲ Thrust faults (sawteeth on upper plate)

⇄ Strike-slip faults (barbs indicate relative motion)

▲⇄▲ Oblique thrust fault

↑ Present motion of Pacific Plate relative to North America, showing rate in inches per year

A L A S K A

NORTH AMERICAN PLATE

Anchorage
Valdez
Seward
Yakutat
Sitka

AM	Aleutian megathrust
BRFS	Border Ranges fault system
CFS	Contact fault system
DFS	Denali fault system
FF	Fairweather fault
QCF	Queen Charlotte fault
TSF	Transition fault system
TCF	Tik clik fault

2 in./yr.

PACIFIC PLATE

2.5 in./yr.

The Tordrillo and Chigmit mountains lie at the northern end of the 1,200-mile-long chain of Aleutian Islands. On a clear day in Anchorage, you can look across Cook Inlet to these snow-covered mountains within the Lake Clark District. They're the real thing: smoking, sometimes belching, stratovolcanoes that sure know how to throw a party. Mount Spurr erupted in 1992. Ash was blown 65,000 feet into the atmosphere and drifted downwind to thinly blanket much of Alaska.

▲ Mt. Redoubt erupted in 1989, instantly melting glacial ice that had accumulated for centuries.

➤ Double Glacier perches on the Chigmit Mountains in front of Redoubt and Iliamna volcanoes, Lake Clark National Park and Preserve.

Redoubt Volcano, 60 miles to the southwest, exploded violently in 1989. Drift Glacier lies on Redoubt's north side. The midriff of this glacier was vaporized by the eruption—almost 400 million cubic yards of ice instantly blown to smithereens. Enormous floods called *lahars*—mushes of melted ice and volcanic ash—swept downhill and seaward, threatening the pipelines that stitch together oil platforms on the west side of Cook Inlet.

Volcanoes are not the only geologic phenomenon known to shake up Alaska. Earthquakes are a fact of life—and death—here. Ask Howard Ulrich. On July 9, 1958, he and his son had just fallen asleep aboard their 38-foot fishing boat, the *Edrie*, in normally quiet Lituya Bay, when they were shaken awake. The Fairweather Range, which rises above the east end of the bay, was writhing and groaning as a Magnitude 8.3 earthquake propagated out from an epicenter just thirteen miles away. Ulrich watched in horror as 40 million cubic yards of rock and ice fell 3,000 feet into the bay. The rockfall spawned a wave that sheared trees 1,740 feet up a nearby wall. The wave raced down the narrow seven-mile-long bay toward the *Edrie*. By the time it reached Ulrich, he estimated its height between 50 and 75 feet, still enough to pitch the *Edrie* nearly vertical as it rode up the face. Ulrich's boat fared better than two others anchored that night in Lituya Bay; both sank, one with the loss of the crew.

The Fairweather Fault underlies Lituya Glacier, running past this western margin of the Fairweather Range. USGS geologists determined that land on the fault's seaward side had jumped vertically three feet and horizontally 21 feet to the northwest relative to the mainland during the 1958 earthquake. Ever since Ralph Tarr and Lawrence Martin explored this coast, scientists have intuited that earthquakes must somehow affect glaciers. Tarr and Martin found evidence of much avalanching of snow onto glaciers in 1899. That earthquake had shaken loose so many icebergs that Glacier Bay was temporarily blocked, its surface described as a "solid mass of floating ice." Photographs taken

immediately before and after the 1958 earthquake show that 1,300 feet of ice were sheared from the terminus of Lituya Glacier by the bay's great wave. But beyond superficial effects such as avalanching and incidental iceberg calving, the glaciers around Lituya Bay did not fundamentally change their behavior due to the earthquakes of 1899 or 1958.

Alaska was rocked by an even greater earthquake, Magnitude 9.2, on Good Friday, 1964, which claimed the lives of 125 people. Thirty-four thousand square miles of south-central Alaska were thrown either skyward by as much as 37 feet, or dropped downward by over seven feet. Anchorage was sliced into tilted ribbons. The water-fronts of Seward and Valdez were largely washed away. A series of tsunamis inundated Kodiak Island to elevations as great as 30 feet above the normal high-tide line. Austin Post raced out to document the earthquake's effect on Alaskan glaciers. He pho-tographed debris flows and snow avalanches—675 million cubic feet of rock had instantly blanketed three square miles of Sherman Glacier near Cordova. Surprisingly, he found no evidence of surges or other tectonically encouraged glacial movement.

In November 2002, a Magnitude 7.9 earthquake shook the Denali Range like a dog thrashing a rag doll. The quake actually tore through the ice of Black Rapids Glacier, but again, there was no indication that the glacier itself sloshed any faster down its valley. There is growing evidence, though, that this cause-and-effect game between glaciers and tectonics may be working in the reverse direction. Folks speculate that earthquakes like Yakutat 1899 and Mount St. Elias 1979 were hastened when nearby

▼ *Lituya Bay was devastated by waves generated by an earthquake and landslide along the Fairweather Fault in 1958.*

ice retreated, taking with it weight that could have previously kept fault planes locked together. So, though earthquakes apparently don't set glaciers careening down mountainsides, glaciers do interact with the Earth through a different mechanism.

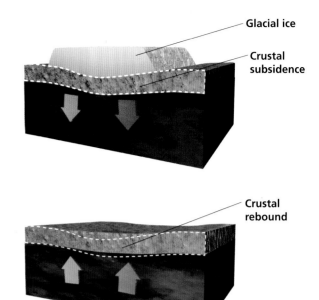

Glacial ice

Crustal subsidence

Crustal rebound

The Earth's crust is a brittle rind, underlain by a thick, viscous layer called the mantle. When sufficient weight is piled up on the surface, the crust bends down elastically and the mantle below oozes viscously. The Laurentide Ice Sheet was heavy enough to depress Hudson Bay by 600 feet during the Pleistocene. When the ice sheet melted, the crust rebounded upward. This rebound has a fast elastic component, as well as a slower visco-elastic component that is still seeking its pre-Pleistocene equilibrium in Canada. The fast component is due to the crust springing back into its preglacial shape, and will be largely accomplished within a couple of hundred years of a glacier's departure. But the slow component may take thousands of years as the underlying mantle slowly squeezes back under the rising crust.

The weight of glaciers can be great enough to depress the Earth's crust. If a glacier subsequently melts, the crust will slowly rebound to its original height.

Not too long ago, Glacier Bay was covered by 230 square miles of ice with an average depth greater than 3,000 feet. Lots of weight. At the end of the Little Ice Age around 1850, Glacier Bay shook off these heavy covers as most of its tidewater glacier system collapsed. Much of southeastern Alaska is now headed skyward. Coastlines at Yakutat to the northwest and Sitka to the southeast are emerging at rates of one- or two-tenths of an inch each year. Juneau is rising by better than one-half inch a year. But Glacier Bay—and especially Bartlett Cove at its mouth—is hurtling up by 1.5 inches each year. You can just about see the salt water streaming down its sides as the land breaches skyward.

The $64,000 question on geologists' minds these days: Is all of this uplift due to glacial retreat? The elastic component of glacial rebound occurs along a curve that decays precisely with time. With accurate sea-level measurements, one could quickly determine if uplift fits the time curve. But the viscoelastic component is a wild card with which scientists are only now beginning to grapple. And as if that weren't enough, there are two other confounding considerations. Glaciers like Margerie and Johns Hopkins are avidly tearing down the very mountains that created them, and the crust is adjusting to the loss of these mountains just as it adjusts to the loss of ice. Finally, the motion of the Pacific Plate undergoes a fundamental change as it sweeps past Glacier Bay. As it glides up from the southeast, the plate slides along the edge of the North American Plate. But passing Glacier Bay, the Pacific Plate begins the terribly complex business of diving back into the mantle. These complications—viscoelastic behavior, continental rebound, and tectonic adjustment to subduction—are likely to keep geologists scratching their heads for years to come.

Kantishna

WILDERNESS

NATIONAL PARK

DENALI PARK ROAD

Eielson
Visitor
Center

Muldrow Glacier

Peters

Straightaway

Foraker

Herron

NATIONAL PARK

Rath Glacier

Tokositna Glacier

Dall

Peters

Lacuna Glacier

Kahiltna Glacier

North

| 0 | 5 | 10 | 15 | 20 miles |

| 0 | 8 | 16 | 24 | 32 km |

Denali National Park and Preserve

Muldrow Glacier in Denali National Park is the largest north-flowing glacier in Alaska. A trip to Denali is the high point of many visitors' Alaskan experience—a dramatic landscape of mountains and ice, braided rivers and rolling tundra, and when the weather is right, breathtaking views of Mt. McKinley, North America's highest point. Denali National Park is accessible by road, rail, or air.

Best Viewing Access

Even when the clouds are too low to view Mt. McKinley, Muldrow Glacier is easily observed from the Eielson Visitor Center deep inside Denali National Park and Preserve. The lower section, closest to the visitor center, is covered in rock that's been deposited on top of the glacier over time. Visitors often don't realize that what they are viewing is actually a glacier.

What Else to Do

Hike the steep climb up to Anderson Pass from Glacier Creek for excellent views of Muldrow Glacier and the Alaska Range.

Book a flight-seeing trip over Mt. McKinley and the Alaska Range in the mountaineering town of Talkeetna. Those wanting an up-close experience can even arrange a stop on Ruth Glacier, from which many mountaineers begin their ascent of McKinley.

Getting There

Denali National Park and Preserve lies in the heart of Alaska, and the Muldrow lies deep in the heart of Denali. The best approach is via one of the shuttle buses run by the park concessionaire. Of note: To get far enough into the park to view Muldrow, you will spend a full day on the bus. Bring all-weather clothing and plenty of food and water. The views of the magnificent Denali landscapes are great from the bus, so don't forget to bring your camera to take snaps of the wildlife you'll see along the road.

Website

http://www.nps.gov/dena

▲ Camping near the Muldrow Glacier (top); flight-seeing over the Alaska Range (bottom).

Chapter Five
THE FRAMEWORK OF ICE

MUCH OF THE WORLD'S MYSTERY can be
found in a single drop of water: two hydrogen and one oxygen
atoms. Oxygen and hydrogen are oppositely charged and attract
each other. But being jealous siblings, the two hydrogens are
partially repelled, careful to keep a discreet 104.5-degree distance
between themselves around the hub of oxygen. H_2O, then, is
nothing but a pointy little ménage à trois. When more than a few
of these molecules congregate, the hyper-social hydrogen atoms
start checking out the nearby oxygens. Chemists call this
"hydrogen bonding." I call it a threat to that cozy arrangement
between oxygen and his harem of hydrogen atoms.

Hydrogen bonding actually turns out to be deliciously significant
in the real world. Without this earthy attraction between
molecules, ice would melt into a liquid at temperatures much
lower than 32 degrees Fahrenheit; water would boil away at
lower temperatures. Rivers would be dry; our bodies would be
dust. Most materials become denser as they cool. But hydrogen
bonds cause adjacent H_2O molecules to be stickier as they chill,
and help to create an interlocked lattice structure as water
freezes into ice. If this were not so, ice would
sink rather than float on water.

Given their druthers, water molecules will
arrange themselves into hexagonal crystals when
freezing. As lake water slowly chills below 32
degrees F, the resulting ice takes the shape of
dense elongate candles. But airborne water
freezes very differently. The atmosphere
sometimes superchills before droplets solidify
into ice. And when water does finally freeze, it
snaps into the shape of a feathery snowflake, a beautiful blossom
rather than a dense hexagonal candle. To be sure, snowflakes are
also hexagonal—after all, they too feel that rush of hydrogen
bonding. But unlike other forms of ice, those lacy snowflakes
have a lot of empty air-filled spaces between their arms and ears.

Snow falls throughout Alaska, but in most places, winter's
accumulation melts away each summer. Snow survival through
a summer–winter cycle requires two conditions: adequate
springtime snowpack and temperatures that infrequently rise

▲ *Calving ice falls into Prince William Sound from
the Beloit Glacier, Chugach National Forest.*

◄ *Mount Barnard above Johns Hopkins Glacier,
Glacier Bay National Park and Preserve.*

◄◄ *Harding Icefield near Seward, Kenai Fjords
National Park.*

above freezing. More snow needs to fall than will melt if we're going to get around to making glaciers.

During their first summer, those feathery snowflakes begin to mature, partially melting and refreezing into sad little blobs of ice, steadily surrendering their original delicate figures. Snow that does survive through a summer is called *firn* as soon as it is buried by the next winter's accumulation. If you recall the last time you shoveled a driveway, you (and your aching back) know that even freshly fallen snow can be heavy—anywhere from three to twelve pounds per cubic foot. Snowpacks that start out 30 (and more) feet deep are quickly compressed into much thinner layers as years of additional snow are piled on top. Firn densities are greater than those of snow—from 25 to 52 lbs/ft³.

Ice crystal metamorphosis.

Snowflakes initially contain about 10 percent ice and 90 percent air. But when the flakes are compressed, most of the air is driven out. As firn collapses even further, the interconnected passageways that allowed free movement of air become blocked, at which point firn finally graduates to full-fledged *ice*. The Suck Test is a valid scientific method of distinguishing firn from ice: Put a cold white block next to your lips. If you can suck air through it, it's firn. If you can't, it's ice. In cold, dry climates, the transformation from snow to firn to ice can take a century or more. The first true ice in a Greenland core might be 200 feet below the surface. But glaciers of coastal Alaska and Canada receive vastly more precipitation and have warm summers that accelerate the transformation to glacial ice. Seward Glacier, above the Malaspina and straddling the border into Canada's Yukon, has ice at depths of only 45 feet, under accumulations of only three to five years of snow.

Pure ice has a theoretical density of 57.25 lbs/ft³, which is about as close as ice gets to the density of water (62.4 lbs/ft³). Glacial ice is slightly less dense than pure ice because it has those air bubbles trapped inside. Glaciers in Alaska can be as much as 3,000 feet thick. Bubbles compressed by this much ice can have pressures in excess of 1,000 pounds per square inch. When glacier movement brings that ice back to the surface, the air bubbles retain the high internal pressures to which they have grown accustomed at depth. Next time you're leaning over the rail on that "756 Glacier Cruise," look up and listen long enough to appreciate that a lot of the little icebergs in the water around you are sizzling. What you are hearing is the sound of air that is hundreds of years old, effervescing out of the ice.

Scientists have been able to make estimates of ancient atmospheric compositions by sampling air trapped in bubbles of glacier ice. These estimates are useful when you're wondering how much more carbon dioxide is in the air now than in the past. (The short answer: lots more.) Glaciers hold other markers of past environments. Minute layers of wind-blown dust are deposited on top of glaciers each summer. Glaciologists can identify the succession of individual years by examining the compressed dust layers in long cores of ice, a process not unlike counting growth rings on a tree trunk. One of the glaciologist's most powerful tools relies on the fact that oxygen comes in two flavors: "regular" and "heavy." That is, ^{16}O and ^{18}O. Each is more or less likely to evaporate from an ocean depending on the ocean's temperature; conversely, each is more or less likely to precipitate out of the sky, depending on atmospheric temperatures.

By sampling glacier ice and measuring the relative amounts of these two forms of oxygen (a ratio called $\delta^{18}O$), glaciologists can sort through clues about ocean and atmosphere temperatures back hundreds and even thousands of years.

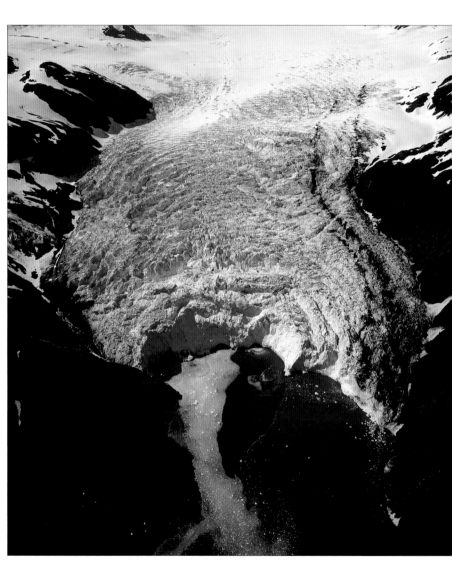

Ice is a crystal, of course. Rocks are collections of crystals. So, technically speaking, rock—i.e., "any naturally formed aggregate of mineral matter constituting an essential and appreciable part of the Earth's crust"—can comprise interlocked crystals of ice. If the shoe fits, wear it. Ice is a rock. Snowflakes and firn grains metamorphose into glacier ice as the boundaries of the smaller grains fuse into larger crystals. Glacial ice crystals can enlarge to the size of a basketball. These crystals transmit light, though that light is curiously modified in its passage. Long-wavelength reds and yellows are absorbed by ice, while blue wavelengths

▲ *Holgate Glacier spills down from the Harding Icefield, Kenai Fjords National Park.*

wriggle through the lattice structure just fine. Thus, though full-spectrum sunlight enters a glacier, only blue light emerges. This light is one of the most sensual aspects of any glacier: a searing blueness that glows from within, no less intense than the burning red that illuminates the throat of a Hawaiian volcano.

▲ *Dennis Trabant commuting to work on Wolverine Glacier, Chugach National Forest.*

➤ *USGS research hut alongside Wolverine Glacier.*

Dennis Trabant and I sat on our snowshoes,

eating lunch, coats cast off in the mid-May sun. Wolverine Glacier hung in the sky above us, brilliant white against the purest blue. Dennis was winding up another year's measurement of the glacier's movement. We had sampled snow pits at progressively lower elevations until we now were sitting a stone's throw from the glacier's snout—"terminus," as he kept correcting me. Down here, the snow was sloppy, melting by the minute. Five miles up-glacier, the previous winter's accumulation exceeded 30 feet. Dennis was ecstatic that so much snow had fallen. I realized that he took this glacier's health very personally. That's bound to happen after coming out here every year for three decades. Being a scientist involves so much more than just gathering numbers. I ate beef jerky and washed it down with fistfuls of slushy snow. We watched without words as a wolf lunged silently through snowdrifts along a ridge a few hundred yards to the west.

▲ *Rod March checking the weather station above Wolverine Glacier.*

Glaciers flow under the influence of gravity. The upper reaches are in mountains where air temperatures are low and snowfall can be high. The lower reaches are bathed in warmer air, with summertime temperatures well above freezing. Tidewater glaciers calve icebergs, a process that enables them to lose mass even faster than by melting. Squint a little: The upper portion of a glacier accumulates more snow than it loses, and the lower portion loses (ablates) more snow than it accumulates. Glaciers have distinct areas of accumulation above and ablation below. These two zones are separated by an equilibrium line where net ice gain exactly matches net ice loss. In late summer, you can pretty much eyeball this equilibrium line—above it, snow still covers a glacier's surface; below it, bare ice lies melting in the sun.

In a static world, glaciers would never change. Additions to the accumulation zone would always exactly match losses in the ablation zone. A glacier would neither thicken nor thin, neither advance nor retreat. But the real world is hardly static. Snowfall varies from year to year as storm tracks migrate back and forth across the region. Summertime temperatures might be high or low depending on the nuances of cloud cover. These changing patterns can either be short-term (weather) or long-term (climate). Climate is what you expect; weather is what you get. We humans, with our relatively short life spans and notoriously short memories, reside in a world of weather. Glaciers exist in the world of climate.

◄ *Colony Glacier, east of Palmer, Chugach National Forest.*

Glaciers receive the most snow at high elevations, while the greatest ice loss occurs at lower elevations. A glacier's upper reaches are therefore areas of accumulation, and its lower reaches are areas of ablation. The two zones are separated by the equilibrium line, where accumulation is exactly balanced by ablation.

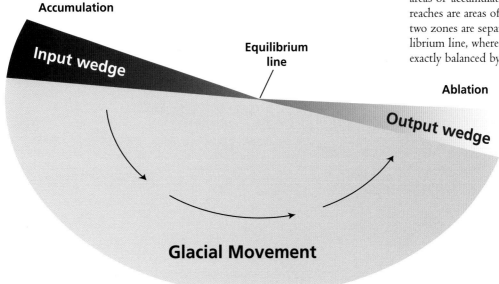

Accumulation

Input wedge

Equilibrium line

Ablation

Output wedge

Glacial Movement

Glaciers do make subtle adjustments to the vagaries of weather—for instance, melting rates might increase during a warm spell. But for the most part, glaciers respond more to changes of climate than weather. A glacier's health can be monitored by watching its equilibrium line altitude move up-glacier or down-glacier from one year to the next. As it does, the ratio between the accumulation and ablation areas will increase or decrease. The accumulation areas of healthy glaciers in coastal Alaska will be 70- to as much as 90 percent of the glacier's total size. Substantial changes of climate alter the shape and motion of glaciers throughout an entire region. Indeed, the Earth's atmosphere has dramatically warmed over the last fifty years and most Alaskan glaciers are now in full-fledged rout.

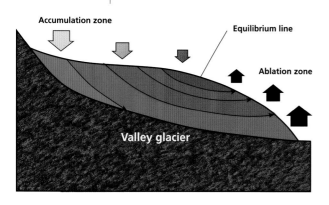

The surface of a glacier is an ever-changing palette. Cracks propagate willy-nilly as though the ice were a shattered windshield. Holes yawn open into unthinkable depths, inviting that final misstep as you lean closer to look. Crazed pinnacles 100 feet tall stand in tilted icy forests. Slithering lines of rock litter the ice like whirling designs in a giant sandpainting. Let's look at some of these features and try to make sense of them.

Taken to reductionist extremes, a glacier is nothing more than a very large grinder and conveyor belt. As ice slides downhill it abrades the bottom and sides of its valley. Fragments varying in size from dust to tour-bus are plucked from the underlying bedrock and entrained in the ice. Rocks on the glacier's bottom will be hidden from view until they are finally dumped onto the terminal moraine down at the snout—excuse me, the terminus. Rocks along the side of a

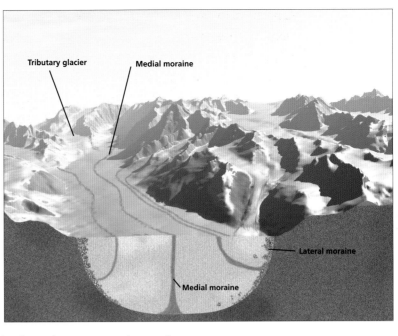

Medial and lateral moraines on a typical valley glacier.

glacier, however, will be exposed for long distances at the surface, precisely at the boundary between the ice and its confining walls. This lateral moraine is a collection formed by rock both pilfered directly from valley walls and falling onto the ice from above. When two valley glaciers converge, their adjacent lateral moraines merge and are carried downstream, out in the middle of the larger glacier, as a medial moraine.

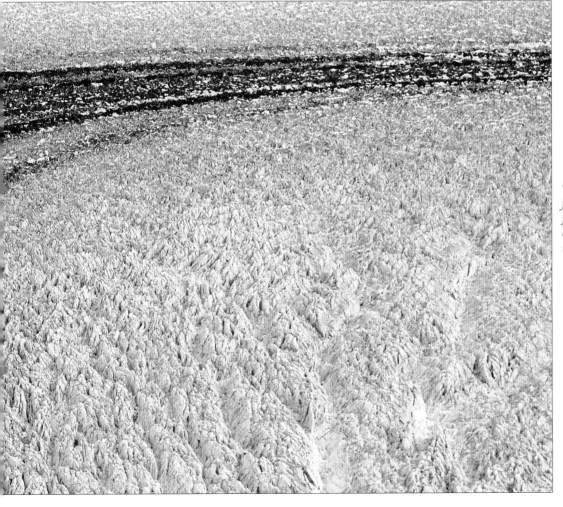

◄ *This medial moraine is formed as Valerie Glacier joins Hubbard Glacier just before arriving at tidewater in Desolation Sound.*

Crevasses are fractures that open when ice is stretched. Tension is created when portions of a glacier speed up and stretch the surrounding ice. As a glacier curves around an obstacle or bend, it is compressed in some places and stretched in others. Cracks help relax the tension that builds up in such a bend. As the glacier continues to flow downhill, a crevasse is carried along as a surface feature on the ice. Soon, tension builds at the bend again and another crack appears just upstream of the prior crevasse. In this way, an entire field of crevasses forms below an area of tension. Upright splinters of ice called *séracs* form where crevasse patterns intersect. When a glacier plummets over a drop in its valley floor, it is pulled longitudinally, with crevasses opening throughout the drop to create the broken maze of an *icefall*. *Ogives* are stretch marks that curve across the belly of a glacier. These alternating dark and light bands occur when ice speeds up and thins out in summertime as it plummets through an icefall. The temporarily stretched ice, with its greater surface area, becomes darker because it collects more airborne dust and debris than adjacent ice that went through the icefall in winter.

Ice at the surface of a glacier stretches very little before it snaps. The warm glaciers of Alaska are only strong enough to support cracks down to about 100 feet. Below this depth, ice is under enough pressure from its own weight to smear out like soft plastic rather than fracture as a brittle material in response to tension. The good news is that if you slip into a crevasse, you aren't likely to fall more than the height of a ten-story building. When a crevasse field migrates downstream and later encounters an area of compression, the crevasses close up and the ice fuses back together.

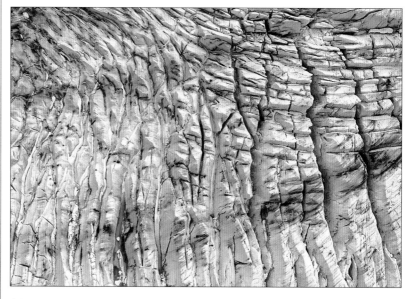

⋏ *Davidson Glacier, near Haines.*

➤ *Crevasses on Crown Glacier, part of the Sargent Icefield above the Nellie Juan River, Chugach National Forest.*

Triumvirate Glacier sweeps down from the Tordrillo Mountains west of Anchorage, temporarily blocking drainage of Strandline Lake. The lake periodically breaks through its ice dam, flooding past Beluga Lake and into Cook Inlet.

Water—produced by melting or rainfall—runs along the top of a glacier and collects into rivulets. These streams tumble either into crevasses or down those vertical tubes called *moulins*, gradually making their way toward the bottom of the glacier. When water does flow at the bottom of a crevasse, its relative warmth can open a larger channel, or even keep that channel open when the rest of the crevasse squeezes shut above it. In this way, internal cavities come to exist within a glacier. The cavities tend to be isolated from each other during the cold winter months because less surface flow is available then. During summer, water flow increases and the conduits enlarge and become interlinked.

Despite the tremendous pressure that necessarily exists at the bottom of an Alaskan glacier, water does manage to percolate all the way down. This water may reside within interconnected bottom conduits or as a microscopically thin layer right at the interface between the ice and its bed. By and large, glaciers in Alaska are not frozen to their beds. This bottom water becomes extremely important when we get around to considering the downstream movement of glaciers.

Water usually flows underneath a glacier peacefully enough, moving steadily toward the terminus. But once in a while, conduits and subglacial streams get gummed up. Sometimes glaciers plow across a canyon and dam an external drainage. Then water begins to pond. Where potential energy accrues, trouble brews. Triumvirate Glacier spills down from the Tordrillo Mountains toward Cook Inlet. A tributary to the north is too low to contribute ice anymore, but it does contribute water. About every three years, Triumvirate replugs this side canyon with a 1,500-foot lobe of ice and Strandline Lake fills. Whenever 25 billion cubic feet of water raise the lake surface to 1,273 feet above sea level, the ice lobe starts to float. Water seeps under the dam and the lake begins to empty downstream. Peak flows exceed 200,000 ft^3/sec—about ten times the typical flow of the Colorado River through Grand Canyon. Within twenty days, Strandline Lake all but vanishes.

Many cataclysmic events go unnoticed in Alaska because the state remains delightfully unpopulated. When Russell Lake occasionally empties past Hubbard Glacier, flow out of the lake can temporarily exceed that of the Mississippi River. A few seals are washed out to sea. A short while after Hidden Creek Lake drains under Kennicott Glacier, the folks in McCarthy notice

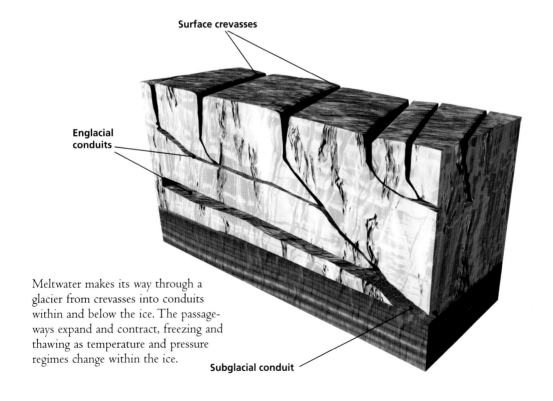

Surface crevasses

Englacial conduits

Meltwater makes its way through a glacier from crevasses into conduits within and below the ice. The passageways expand and contract, freezing and thawing as temperature and pressure regimes change within the ice.

Subglacial conduit

that the Kennicott River has risen, but not to levels likely to threaten their footbridge. People might take these floods more seriously if they could pronounce the Icelandic name by which they are known: *jökulhlaups*. Periodic floods from Strandline Lake, however, do undermine bridges and major powerlines that serve Anchorage. The Chugach Electric Company has enlisted glaciologists to study the floods, but so far, they have accomplished a lot more mustache-twirling than preventive engineering.

So we've looked at some of the details of glacial structure. Let's now turn our attention to how glaciers move.

▲ *The 1964 Good Friday earthquake violently shook all of southern Alaska. East of Cordova, in the Chugach National Forest, a great mass of debris was knocked loose from Shattered Peak. The rock blanketed the lower reaches of Sherman Glacier, and actually protected that part of the glacier from melting.*

◄ *Rockfalls that occurred in the mid-1990s within Denali National Park and Preserve now ride Buckskin Glacier downhill. The lobes gradually change shape due to faster movement at the center of the glacier.*

Chugach National Forest

Portage was once one of Alaska's most easily viewed glaciers; though its rapid recession has left a large lake in its wake, Portage still offers easy access to glacier tales.

Best Viewing Access

Go around the bend—a boat tour of the lake provides up-close glacier views. The glacier can also be seen on the road to Whittier to/from Portage before entering the tunnel.

Take an uphill hike from the Whittier side of the Whittier Tunnel to Portage Pass for breathtaking views of Portage Glacier and the surrounding terrain.

◄ Iceberg on Portage Lake, with hoary frost of winter (top); Looking up Portage Creek valley (middle); Begich, Boggs Visitor Center in autumn (bottom).

What Else to Do

Visit the Begich, Boggs Visitor Center, which offers award-winning hands-on exhibits. Short hikes abound in Portage Valley, including one that goes to Byron Glacier; trailhead begins just 2 miles from the visitor center.

Getting There

Portage is about an hour's drive south of Anchorage on the Seward Highway, which skirts scenic Turnagain Arm.

Website

http://www.fs.fed.us/r10/chugach /chugach_pages/bbvc.html

ALASKA RAILROAD

2780'x

Portage Creek

TUNNEL

Bear

Williwaw CG

Gary Williams Moraine Trail

Begich-Boggs Visitor Center

PORTAGE LAKE

Portage Glacier Tour

Byron Glacier Trail

I waved as the Skywagon thundered away,

rising out of its cloud of propeller-whipped snow. A meaningless gesture—

the pilot surely wasn't watching, and there wasn't anyone else around to

notice. The ski-plane disappeared down Kahiltna Glacier. Echos of the engine

gradually died away like ripples subsiding on a windless pond. I was deep in

the Alaska Range. To my left, ice loosened by the mid-day sun hurtled down

the face of Mount Hunter. Behind and to the right, Denali loomed two and a

half miles overhead.

Alone, I didn't care to walk far. The year's snowcover was gone and a

maze of open crevasses criss-crossed the glacier's surface. If any snowbridges

remained to conceal crevasses, they would be as firm as powdered sugar.

I sat at my tent's door, sometimes reading, mostly just watching the moun-

tains. A storm rolled in and plugged up the sky for days. I huddled atop this

southeast fork of Kahiltna Glacier, waiting for the plane to return. I felt more

than heard the deep slow groans of ice beneath me. I had metamorphosed

into a vanishingly small organism riding the back of a great sleeping creature.

▲ *A Cessna 185 at the Denali base camp above Kahiltna Glacier.*

➤ *The flanks of Mt. McKinley, part of the Alaska Range.*

◄◄ *Great Gorge, Ruth Glacier, Denali National Park and Preserve.*

ICE IN ALASKA is a solid that exists quite near its melting point. Cold Antarctic ice, separated from its melting point by a few additional degrees, behaves as a brittle substance to deeper depths. But here, ice is softer and more malleable, a little more willing to flow. Mendenhall Glacier probably weighs something like 50,000,000,000,000 pounds—give or take a few trillion. It's tempting to ask how such a tremendous mass could possibly move. But that would be the wrong question. Better to ask, what could *prevent* it from moving? The inexorable movement of glaciers in Alaska has carried many a mountain down to the sea.

Scientists like Tad Pfeffer and Neil Humphrey from universities in Colorado and Wyoming have painstakingly studied the movement of Worthington Glacier above Valdez. Midway up, they found average surface velocities of about 270 feet per year. They showed that the top of the ice always travels faster than the bottom. The middle of the glacier always flows faster than its edges. The glacier moves faster in summer than in winter. The glacier can speed up and then slow down on a daily cycle. Depending on the balance between rates of accumulation, movement, and ablation, the terminus of a glacier like Worthington can advance or retreat. That's what it looks like—ribbons of ice careening around Alaska like a motley crowd of misbehaving snakes. But how does it all work?

Glaciers sail downhill by one or a combination of three mechanisms. Ice can slide on its base, bumping along as a solid mass. The base can deform, transporting the intact glacier as though it were being carried on a moving sidewalk. Or the glacier can internally deform, spreading itself out like silly putty on a hot sidewalk.

OK, one at a time. Basal sliding is the most intuitively accessible of the three mechanisms. After all, everyone knows that ice is slippery—remember the last time your feet flew out from beneath you, as, arms windmilling, you tried to walk

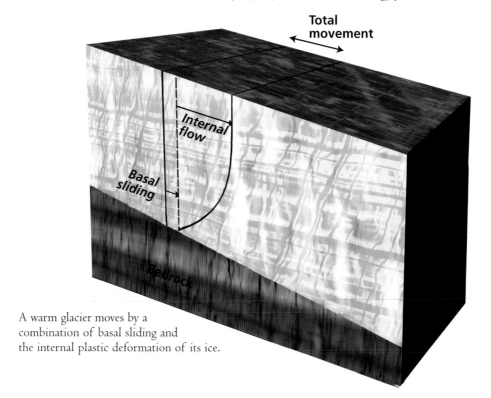

A warm glacier moves by a combination of basal sliding and the internal plastic deformation of its ice.

on an icy street? Now sprinkle a little water on the street and try to walk on it. Good luck. Alaskan glaciers have a fine film of water lubricating their bases. Sliding is enhanced when basal water pressure is increased. Hydraulic jacking occurs when a summertime rainstorm dumps water into a well-developed conduit system and actually lifts the glacier up off its bed.

Again, the question arises: Why don't they all just zoom out to sea? For the answer, let's examine parts of Mendenhall Glacier's bed that have been exposed by its recent retreat. The bedrock is schist and gneiss that is exposed below the westside trail. It has been etched with fine parallel grooves, gouged by rocks that once protruded from the very bottom of the glacier. The bedrock is hummocky; it would have presented many large barriers to the ice as it slid downward. Roughness, caused by sand grains and gravel impregnated in the bottom of the ice, dramatically increases friction between a glacier and its bed.

Ice that encounters a rough bed has a few tricks up its sleeve. You may think that water freezes at precisely 32 degrees F. But that is only true at "standard" pressure. Increase the pressure (as occurs on the upstream side of an obstacle), and the freeze/thaw temperature drops and the ice melts. Conversely, lower the pressure downstream of the obstacle and the freeze/thaw boundary rises and water refreezes. Even when the temperature remains constant, glacier ice is able to melt above and refreeze below an obstacle. You would be performing the same trick if you walked through a wall instead of around it.

Now, bed deformation. Very often, glaciers sit not on clean hard bedrock, but on gooey layers of ground-up rock called *till*. Think of till as a wet concrete slurry with varying amounts of aggregate, sand, and water. It slumps under its own

▼ *Packed icebergs in Johns Hopkins Inlet, Glacier Bay National Park and Preserve.*

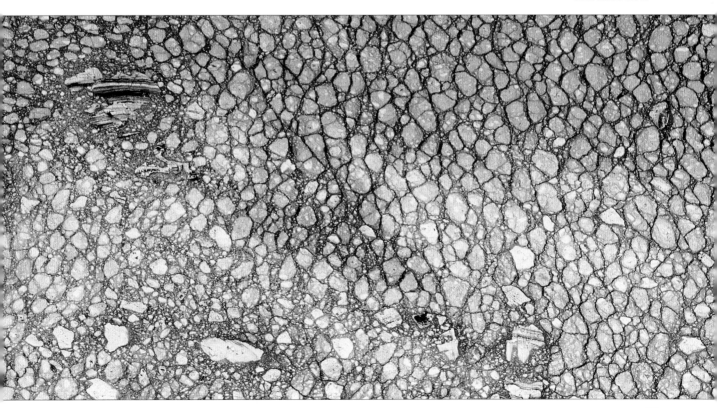

▲ *Alaska Range near Black Rapids Glacier.*

weight, let alone under the weight of an overlying glacier. Till does have some strength, determined by its cohesiveness and the internal friction of grains rubbing against each other. But above some critical shear stress, till will begin to flow downhill, carrying its glacier along for the ride.

Folks from the University of Alaska in Fairbanks have spent a lot of time in the central Alaska Range studying the Black Rapids Glacier. These scientists have calculated that the surface is galloping along at 200 feet per year. They drilled through 2,000 feet of ice and found 22 feet of water-saturated till beneath the glacier. The drillholes were spangled with an expensive array of piezometers, pressure transducers, and tilt-meters. They figured out that about one-third to one-half of the glacier's total movement was accomplished by internal deformation of the ice (that's our third mechanism; more about that in just a second), but a full half to two-thirds of the motion occurred on a horizon a few yards into the till layer. Almost no motion occurred at the till/ice boundary. Black Rapids Glacier straddles the Denali Fault for 25 miles. Perhaps the fault premasticates rocks before the glacier gets the chance to further grind them up. This till-dominated movement isn't likely to be representative of all Alaskan glaciers, but it does point out the potential importance of bed deformation when till is present.

The third mechanism of glacier motion—internal deformation—gives me (and glaciers) the creeps. It happens mysteriously down inside a glacier, hidden from all but a physicist's prying eyes and equations. We know that if a shear stress is applied slowly, ice will behave like warm taffy—flowing plastically. This movement occurs crystal by crystal. The plastic flow, or *creep*, of any one crystal accounts for only the tiniest amount of forward motion. But add up the motion of a few quintillion crystals and the glacier starts to build a real head of steam.

Internal deformation takes us back to those hexagonal ice crystals. Any solid (ice, metal, you name it) that is close to its melting point can relieve shear stress as crystals glide along structurally weaker surfaces called *cleavage planes*. The preferred cleavage plane of ice is parallel to the surface on which that six-sided snowflake first came to rest. Individual crystals are continuously growing and shrinking within a glacier. The ones that are favorably oriented to the current stress field will grow at the expense of others that aren't. An ice crystal will shear somewhere between ten and a thousand times easier in a plane

▼ *Blackstone Glacier above Prince William Sound, Chugach National Forest.*

parallel with (rather than perpendicular to) its cleavage plane. This process of work-softening is then responsible for a portion of the ice's plastic flow. Creep also occurs with the migration of molecular defects. These wrinkles progressively shift a crystal's interior layers down the stress field. Again, no one crystal makes much progress individually, but the process is very effective when added up throughout a glacier.

When too much stress is applied to ice too quickly, the patient processes of creep give way to large-scale fractures. Crevasses open up when ice fails under tension. Thrust faults tear through a glacier at depth when the ice fails under compression. These fractures temporarily relieve enough stress that the slower processes of creep can subsequently resume.

⋏ *Crevasses and meltwater ponds on Meade Glacier, Tongass National Forest.*

➤ *Moraines dissected by crevasses on Lamplugh Glacier, Glacier Bay National Park and Preserve.*

Full flaps, 55 knots indicated,

settling at a couple of hundred feet per minute. My plane's windshield is completely filled by the grinning six-mile-wide face of Taku Glacier. We cross the shore of the Taku River and touch down. The landing is soundless, not because of any particular skill on my part, but because we are floating along on a carpet of moss. I kill the engine and we're engulfed in perfect silence. Juneau is only 35 miles away as the buzzard flies, but it might as well be in another universe if we can't get the engine restarted.

Taku is the only glacier flowing from the Juneau Icefield that is advancing; the other thirty-seven are all retreating. Landing here one hundred years ago, we would have sunk because at that time, this was a deep bay. Since then, sediment carried down the Taku River has stabilized the toe of the Taku. And the 4,800-foot-thick glacier has plowed forward, pushing up the crumpled ridges of rock and sand that surround us. At the rate it is going, Taku Glacier could collide with the opposite bank of the river, again damming the Taku River and creating a great lake not unlike one described in Tlingit legends.

➤ *In recent times, Taku Glacier has been the only major advancing glacier to flow from the Juneau Icefield, Tongass National Forest.*

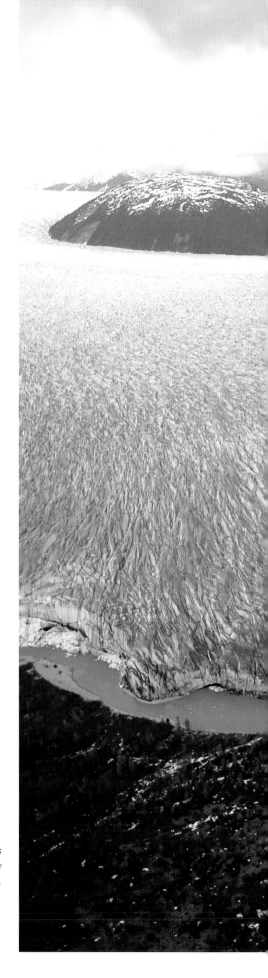

▲ Icebergs from Columbia Glacier in the Chugach National Forest became stranded on its old terminal moraine during the dramatic retreat that began in 1984.

Glaciers are forever poised in tenuous balance between freezing and melting. Snowfall fluctuates year by year. Seasonal atmospheric temperatures might rise or fall. More or less water flows into and under them. These variables in turn influence the three processes that allow glaciers to move in the first place (basal sliding, bed deformation, and internal deformation). To complicate matters further, a glacier responds not only to recent events, which can alter its behavior quickly, but also to events that occurred ten, fifty, or two hundred years ago. Cores from Greenland reveal a clear record of fluctuating temperatures at which snow was deposited thousands of years ago. With so many competing variables, it would be astonishing if glaciers maintained consistent sizes and shapes. In fact, they rarely do for very long.

It's as natural for a glacier to advance and retreat as it is for you and I to breathe in and out. If the balance between accumulation and ablation is positive, a glacier will either grow thicker or longer or both. If the balance is negative, the glacier will lose volume, either by thinning or retreating. Despite connotations of the name, retreat never involves ice flowing back uphill. To make that happen, we would have to rewrite the laws of physics. Isaac Newton would not be happy.

Advancing tidewater glacier

Thinning decreases the supply of ice available to be sent down-glacier. Retreat occurs when melting or calving occurs faster than ice flows forward. Curiously, the two processes are not always coupled. For example, the terminus of Worthington Glacier, like so many throughout Alaska, has steadily retreated in recent years. But during the early 1990s, Worthington was simultaneously becoming thicker and shorter; its total volume increased while its length decreased. These processes are able to operate seemingly at cross-purposes because each affects a glacier at different elevations or on different timescales. Snowfall exerts its immediate influence in the upper accumulation zone. Atmospheric warming increases ablation at lower elevations, where ice is already on the verge of melting.

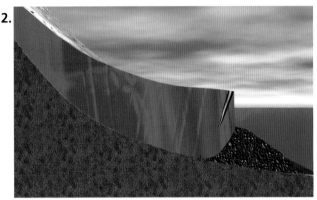

Fully advanced stable position

Terminal moraine shoal

Retreating tidewater glacier

Tidewater glaciers advance and retreat as well, but they march to a different drummer than other Alaskan glaciers. Calving an iceberg is a very efficient means by which a glacier rids itself of down-valley ice—much more efficient than melting terminal ice. A tidewater glacier has only to kiss its bergs goodbye and let the warmer ocean water do the melting for it. In fact, calving is so efficient that a tidewater glacier would not be able to extend into a fjord if it weren't for the protection afforded by a second balancing process: A tidewater glacier in Alaska always builds an underwater berm that protects its terminus. The calving face of a tidewater glacier rarely extends more than a few hundred feet below sea level, even though the associated fjord may be thousands of feet deep. Glacial ice behind the berm can be a couple of thousand feet thick—after all, that's what carved the fjord in the first place. But the glacier maintains a sheltering berm of boulders and mud between its seaward face and deep open water.

When, for whatever reason, a tidewater glacier retreats from its protective berm, calving suddenly accelerates in an unstable fashion. Perhaps this retreat from the berm occurred because the glacier's supply of rock suddenly slowed down. Or the glacier extended too far into a warm low-elevation environment and the balance between accumulation and ablation was thrown off-kilter. Perhaps the glacier just ran out of luck. In any event, the pulse of calving quickens as icebergs are able to tear away from a greatly increased underwater face of exposed ice. Before you know it, the glacier is in full-blown retreat.

The advance of a tidewater glacier, then, is at least partially controlled by its ability to scour out and deliver rock debris to its terminus. Tidewater glaciers have been observed to steadily roll their terminal shoals seaward at about 100 feet a year. Austin Post and Mark Meier (also with the USGS) looked at thirty-three fjords around southeastern Alaska. Moraines in the fjords averaged nineteen miles between their most advanced and most retracted positions. So Post and Meier concluded that tidewater glaciers go through an advance-and-retreat cycle about once every thousand years. Subsequent study of glaciers in Disenchantment Bay and College Fjord suggests that the advance-and-retreat cycle may take even longer to accomplish. This cycle seems to occur over and above other factors affecting a glacier's immediate health, including major climate shifts.

While most glaciers in Alaska have been retreating, some tidewater glaciers have been advancing. This paradox exists because the natural cycle of a tidewater glacier is to slowly advance for many hundreds or even thousands of years, and then quickly retreat. If the natural ratio of time spent advancing relative to retreating isn't thrown off balance by outside forces, something like nine tidewater glaciers ought to be slowly advancing for every one that is retreating.

Post and Meier noticed that the previously stable Columbia Glacier retreated from its shoal along Heather Island in late 1978. By 1980, they were predicting a large-scale collapse. Not much happened until 1984, when Columbia Glacier *did* start to catastrophically retreat. The terminus raced back one mile in the first summer and continued to retreat about one-third of a mile each year well into the 1990s. The main stem of Columbia Glacier sped up, achieving down-valley surface velocities of 100 feet per day at its lowest reaches. Many icebergs were trapped between the calving face and the shallow terminal moraine near Heather Island, but enough made their way into Prince William Sound to wreak havoc. The supertanker *Exxon Valdez* ran aground on Bligh Reef as it attempted to dodge icebergs from Columbia Glacier in 1989.

Black Rapids Roadhouse still stands alongside the Richardson Highway between Delta and Paxson. Abandoned, but standing. The muddy Delta River rises in the

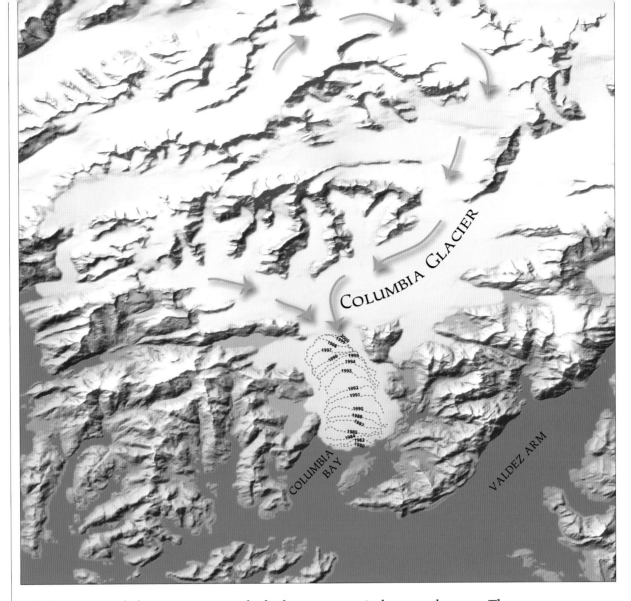

Alaska Range and churns just across the highway, a stone's throw to the west. The roadhouse windows are boarded up now. The walls are decorated with optimistic FOR SALE signs. Others more realistically assert NO TRESPASSING. Before the Parks Highway was built near Denali, this roadhouse straddled the best (read "only") way to drive from Fairbanks to Anchorage. What could have been finer on a crisp morning in November 1936 than to step outside, hands warmed by the cup of coffee that Mrs. Revell the caretaker had given you, and look across the river to the foot of Black Rapids Glacier? That big-barreled block of ice was aimed squarely at the roadhouse, but, shoot, its terminus was miles away.

Then, in the winter of 1936/37, the previously somnolent Black Rapids Glacier charged down from the Alaska Range, covering the three miles that had separated it from Black Rapids Roadhouse in three short months. The glacier threatened the highway but eventually stopped just one-half mile short of the Delta River. It was a false charge. The owners of the roadhouse left the area, and the roadhouse slipped into disrepair. Other structures like the Trans-Alaska Pipeline have since moved into the neighborhood. Fortunately, the USGS succeeded in talking engineers into locating the pipeline farther away from Black Rapids Glacier than originally planned. As Mrs. Revell once learned, it's wise to let sleeping glaciers lie.

Any glacier can vary its velocity a little bit. The lower reaches of a tidewater glacier will speed up and slow down by 10 percent as the tide goes out and comes back in. A valley glacier seasonally accelerates as summertime rainfall pumps more water to its base. A warm Föhn wind can measurably increase a glacier's rate of surface melting and, consequently, its rate of basal sliding. These variations give us glimpses of the little quirks of glacier behavior. But surging glaciers like Black Rapids operate in an entirely different league. Their velocities can increase ten-fold for months at a time.

Austin Post had catalogued 204 surging glaciers in Alaska and western Canada by 1969. More have been added to his list since then. Post combed his aerial photographs

➤ *Hubbard Glacier advanced far enough in 2002 to effectively block the normal drainage of Russell Fjord into Disenchantment Bay. In the past, similar blockage has raised the fjord's level sufficiently to force drainage at its southern end, threatening nearby Yakutat.*

for evidence of surging—the sudden appearance of chaotic crevasse patterns, contorted moraines, and stranded ice in a glacier's deflated upper reaches. As Post analyzed many of these glaciers, he realized that glacial surges were not haphazard events wracking random glaciers on a whim. Surges occur at tantalizingly predictable intervals on particular glaciers.

Variegated Glacier sashays toward Russell Fjord just above Disenchantment Bay. This glacier was known to have surged in 1906, sometime in the late 1920s or early '30s, in 1947, and during the winter of 1964/65—intervals averaging a bit less than twenty years. So glaciologists turned out in full force to study Variegated Glacier for the surge that was due in the early 1980s. They were not disappointed.

▲ *Variegated Glacier flowing toward Russell Fjord, Tongass National Forest.*

In January 1982, seismometers began to pick up a flurry of "icequakes" heralding some sort of internal jostling within the ice. The upper glacier gradually sped up, reaching 34 feet per day in June, then abruptly slowed. A second phase began in October, again announced by increased seismicity. This time, the motion of Variegated Glacier peaked at almost 50 feet per day in mid-June 1983, and again abruptly slowed soon thereafter.

The glaciologists were fascinated by the fact that water almost ceased to flow out the bottom of Variegated Glacier during each of its two surge phases. And then, precisely as each phase ended, a tremendous flood of muddy water erupted from beneath the glacier's terminus. The scientists drilled holes through the ice and discovered that water pressures rose dramatically during the surge and fell precipitously as it ended. Differential motion within the boreholes confirmed that 95 percent of the surge motion was due to increased basal sliding. They concluded that subglacial flow had been ponded and pressurized in poorly connected chambers, thus lifting the glacier up off its bed to slide more easily during the surge.

Surges certainly offer a dramatic view of seemingly unstable processes in nature. After the 1982/83 surge, the upper reaches of Variegated Glacier were deflated by 150 feet, while the lower reaches were inflated by as much as 300 feet. Perhaps this staccato behavior is Variegated's way of adjusting to the vicissitudes of local climate. The glacier may have to muster its reservoir of upper-glacier ice to some threshold beyond which the velocity of ice flow geometrically accelerates. Or maybe there is something about a glacier's bed and subglacial conduits that can temporarily corral basal water flow to initiate the surge. Post and others tried unsuccessfully to estimate which glaciers should surge based on their locations and surrounding geology. Surges are yet another intriguing aspect of glaciers, answering one question even as science raises another.

Tongass National Forest

Flowing from the Juneau Icefield, Mendenhall Glacier drops out of the mountains to the edge of one of Juneau's residential areas. This landlocked glacier has retreated more than a mile since the 1940s. Nearby school children call the Mendenhall their Backyard Glacier, since the school and playground were once below the glacier ice. Open year-round, Mendenhall Glacier Visitor Center, operated by the Forest Service, offers a glacier observatory, a short film, and interpretive displays and talks.

Best Viewing Access

The Mendenhall Glacier Visitor Center is situated at the glacier's terminus overlooking a lake created by and named for the glacier.

◄ Mendenhall Glacier.

▼ Ice at the terminus of Mendenhall Glacier.

What Else to Do

Make a quick stop at the visitor kiosk for information on the area's many trails—from easy walks to steep climbs—surrounding Mendenhall Glacier. Twice a day in summer, rangers lead one-hour interpretive walks.

Getting There

Only 12 miles from downtown Juneau, Mendenhall Glacier may be the only glacier in Alaska (or anywhere) that can be reached via a city bus route and a short walk. Those with their own wheels should follow the Glacier Highway to the Mendenhall Loop Road, which eventually ends at the visitor center.

Website

http://www.fs.fed.us/r10/tongass/districts/mendenhall

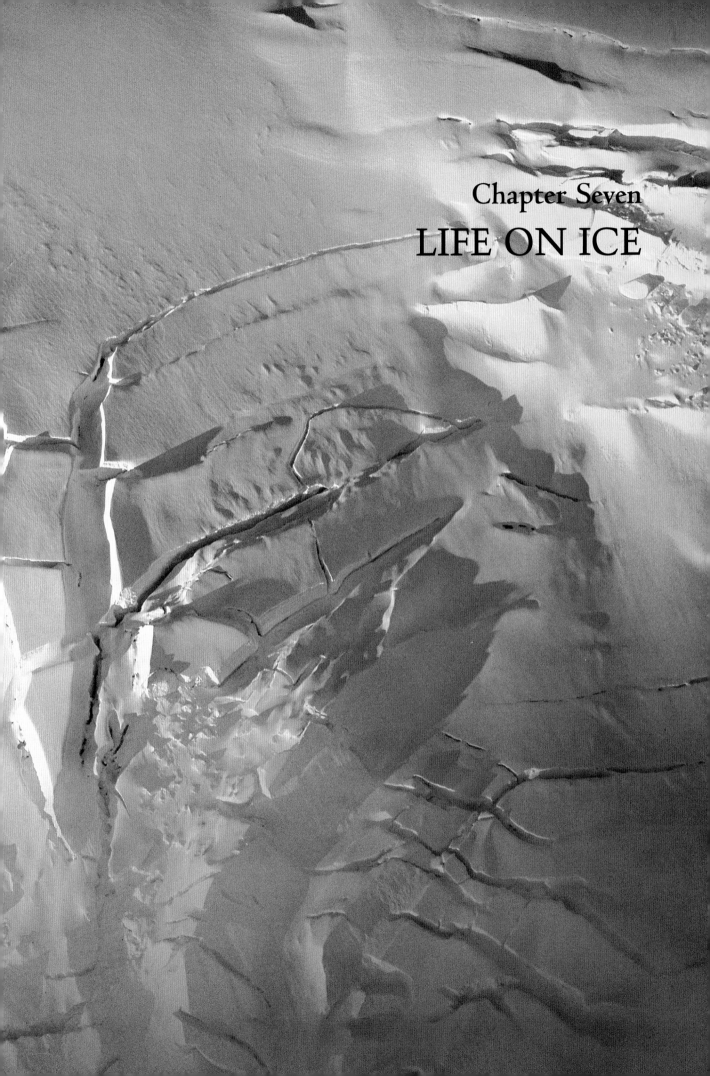

Chapter Seven
LIFE ON ICE

Rumors of a particularly pesky bear

had driven us farther than we'd planned into Tarr Inlet last night. This morning, breakfast done and the kayaks loaded, we shoved off just in time to see Mr. Grizz ambling toward what had been our camp. He shuffled along the shore, taking time to happily rub his rump on a beached block of ice. So much for our efforts to avoid such an encounter. We cruised north, deeper into Glacier Bay. The two boats were a pretty sight, each with two paddlers pulling in unison through the water. By noon we were at our next camp, a 50-foot knob that overlooked the combined fronts of the Margerie and Grand Pacific glaciers. The glaciers had retreated recently enough that only fireweed and alders grew here. Within sneezing distance of our tent I found fresh tracks of moose, wolves, and. . . . bears. Here we go again.

Three nights before, my wife Rose and I had camped just down the shore from Bartlett Cove. Walking along the trail, we had exchanged nervous glances as we gingerly stepped over impressive steaming clues that

▲ *Grizzly on the shore of Tarr Inlet, Glacier Bay National Park and Preserve.*

we shared this forest with our ursine brethren. We pitched our tent on a rock perched at the edge of open water. Around midnight, Rose urgently shook me awake, stage-whispering that something huge was breathing just outside our tent. I suggested that she go out and see what it was. That didn't fly. I slowly unzipped the door and crawled outside into the soft twilight that passes for a summer's night in this part of Alaska. My lower lip quivered. I looked around and then started to laugh, wiping seaspray from my face. A humpback whale was less than 40 feet away, riding a gentle swell, rhythmically blowing and breathing.

Philip Hooge, a USGS biologist, lives with his wife aboard their sailboat, which is anchored in Bartlett Cove. He was cruising around Glacier Bay not too long ago in the *Davidson*, a research vessel tricked out with sidescan sonar and multibeam echosounders. Phil and his partners were studying the erosional activity that gnaws down mountains around Glacier Bay. It doesn't take a rocket scientist—or even a geologist—to guess that the bay's bottom must be deeply buried under rocks and a lot of mud. But those aboard the *Davidson* were not prepared for what they saw. The bottom of Glacier Bay is wildly gouged. What could have scratched these lines? Not glaciers—their bulk would have left very different tracks.

When Vancouver looked toward Glacier Bay from Icy Strait in the late 1700s, all he saw was a great cliff of calving ice. Behind the face, the main glacier trunk was

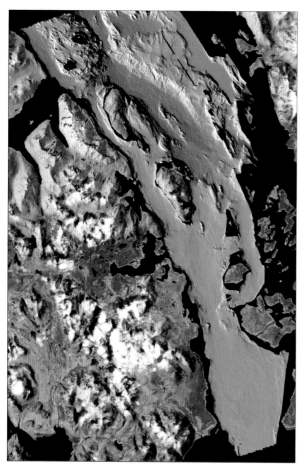

3,000 feet thick. Eighty-five years later, John Muir somehow managed to talk his Tlingit guides into pressing 30 miles up the bay despite miserable weather and too many icebergs. Between the excursions of Vancouver and Muir, the ice of Glacier Bay had receded dramatically. Icebergs, some 500 feet tall, crashed down from the retreating face and sailed off toward the sea. And as they moved, the keels of the greatest bergs dragged through the mud on the bay's bottom.

Phil Hooge has pored over the sonar scans, fascinated by this unexpected diary of iceberg migration. The tracks comb the bottom, generally trending down-bay. But some tracks wander willy-nilly, a record of iceberg peregrinations driven by changes in the wind, odd tidal currents, or maybe just iceberg curiosity. Being not only a biologist but also a scuba-diver, Hooge has gone down to take a look. The gouges are three to six feet deep, and fifteen to 160 feet wide; some continue

Computer generated image of the floor of Glacier Bay, from USGS website.

uninterrupted for five miles. And they are home to a remarkable community of animals that apparently wouldn't be here otherwise. Hooge found twice as many species of fish and sessile (permanently attached) creatures in the gouges as on nearby ungouged areas.

Hooge's work with benthic communities in Glacier Bay precisely reflects the reason this land was set aside as a research refuge in the first place. Early scientists saw Glacier Bay as a grand outdoor laboratory perfectly arranged to study the evolution of a newly revealed terrain. Deglaciation leaves a clean slate: bare rock that must be broken down into soil, an initially sterile environment that only gradually acquires nutrients, and a complex web of life. Ecologist William S. Cooper's dream of having Glacier Bay designated a national monument bore fruit in 1925; the monument was updated to national park status in 1980.

Cooper studied here in 1916, returning in 1921, 1929, and 1935. He found that plants marched up-bay in a tightly prescribed successional cadence as the ice retreated. Pioneers on barren ground include moss, fireweed, and horsetail reeds. The tiny seeds of the fireweed are easily carried on the wind, and the plants reach reproductive maturity at startlingly young ages. Within a quarter-century of deglaciation, a dainty flowering plant called dryas spreads in thick mats over what

Stages of ice reces-
sion in Glacier Bay.

had been bare ground. A few decades later the land is dotted with willows and covered with alder thickets. Both dryas and alders are exceptional nitrogen-fixers, providing a nutrient needed by the plants that will follow them. Sitka spruce and a few cottonwoods manage to rise above the alders within a century of deglaciation. Eventually, western hemlock elbows its way into the spruce canopy; together they share center stage as a climax community.

Ice margins withdrew into the far reaches of Glacier Bay in a more-or-less orderly retreat. As a result, the landscape's adaptation to change can be analyzed as a series of geographically organized slices of time. Moving up-bay toward the current tidewater faces of Muir and Grand Pacific glaciers, progressively younger vegetation and plant communities are encountered. In 1973, Greg Streveler counted 167 rings on a Sitka spruce at Bartlett Cove—that tree would have sprouted twelve years after Vancouver sailed up Ice Straits in 1794. Streveler's spruce was ten miles farther down-bay than one found by William Cooper at Bear Track Cove; it germinated in 1845. Of course you'd want to look at more than two trees, but the ones mentioned by Streveler and Cooper do illustrate the up-bay march of species at Glacier Bay.

Treated as a static concept, a "climax community" assumes that plants evolve toward a predetermined final equilibrium. That idea has come under fire recently as botanists look at the bigger picture of species interaction and global climate change. A bark beetle has recently decimated Alaska's Sitka spruce—perhaps a hundred million trees had died by 2002. Various species of bark beetle have lived with the spruce for eons but a new species was introduced in the early 1990s. This particular

beetle's reproduction rate spiraled out of control as the climate warmed in the last years of the twentieth century. The decimation peaked in 1996 and has slowed down somewhat since then. It's too soon to say how far the spruce and hemlock have been thrown out of balance.

Glaciers also upset the balance of forests. Advancing glaciers like the Taku regularly bulldoze trees that stand in their way. When glaciers retreat, they commonly leave a telltale trimline, with trees above and bare rock below this line marking the ice's highest extent. Great concentrations of glacial till accumulate on the lower ablating surface of large glaciers like Malaspina and Grand Plateau on the Yakutat coastline. These traveling nurseries harbor their own successional micro-ecosystems as lichens and then soils develop on top of the remaining ice. They are crowned with conifers that tilt and sway as the underlying ice melts away. Eventually, the itinerant forest is dumped unceremoniously at the glacier's terminus.

▲ *Sea lion, Aialik Bay, Kenai Fjords National Park.*

▼ *Sea lions at the mouth of Resurrection Bay, Kenai Fjords National Park.*

Few organisms manage to live their entire life cycles directly on glacial ice. Red algae can bloom across great expanses of ice in summertime. Snow fleas and glacier worms graze on the algae or on conifer pollen carried in by the wind. The glacier worm is happiest when its surroundings are right at 32 degrees F and not at all sunny. This three-quarter-inch, thread-like animal will crawl up to the glacier's surface briefly at sunset if conditions are favorable. The snow flea, no bigger than the head of a pin, distinguishes itself by doing a sprite-ly jig on top of the ice.

A number of mammals spend short portions of their lives on or around glaciers. Small icebergs and innumerable bergy bits clutter the surface of Johns Hopkins Inlet, where thousands of harbor seals nurse their pups during summer months. Bears are sometimes seen traversing glaciers, though they rarely have good reason to spend too much time on the ice. By and large, glaciers are tough places to make a living.

In his book *Travels in Alaska*, John Muir described several hair-raising experiences he had while exploring glaciers in 1880. On one occasion, Muir and the little dog Stickeen were faced with a giant crevasse and the very worst "sliver of an ice bridge" Muir had ever seen. The ends of the down-curving, seventy-five-foot-long sliver were attached to the sides at a depth of about eight or ten feet below the glacier's surface. Cutting footholds in the glacier's side and then sliding across the narrow ice bridge, he flattened a 4-inch path for Stickeen to follow. Man and dog thus crossed the crevasse, but the experience was not one Muir wanted to repeat.

Glaciers continue to be fascinating places to explore, but danger lurks—it takes very little time to perish from hypothermia in the frigid grip of a crevasse.

Some general safety tips:

•First and foremost, observe and obey warning signs and exercise common sense.

•You don't have to be touching or standing on a glacier to be at risk. Glaciers are, by definition, unstable. Towering pinnacles of ice called seracs can fall without warning; blocks of ice can tumble down from icefalls. It is inadvisable to enter ice-caves or to stand near steep walls of ice. If you see pieces of broken ice lying on the ground, keep your distance.

•Glaciers' smooth, apparently continuous surfaces are enticing, but walking on them is not as easy as it might seem, and shouldn't be attempted without advance instruction and the proper equipment.

•Technique aside, walking on snow-covered glaciers exposes hikers to the dangers of hidden crevasses, which can be deep and wide, and of tenuously balanced snow bridges, which can collapse under a hiker's added weight.

•Kayaking among glaciers opens a wide new world of adventure—and provides many opportunities for harm. Ice that breaks loose beneath the water's surface can leap skyward well away from a tidewater glacier's face. Glaciers calve unexpectedly, throwing up walls of broken water that, every year, swamp even the most experienced boaters. Best to start this avocation slowly, and ideally with a guide.

◄ *Bear Glacier, Kenai Fjords National Park.*

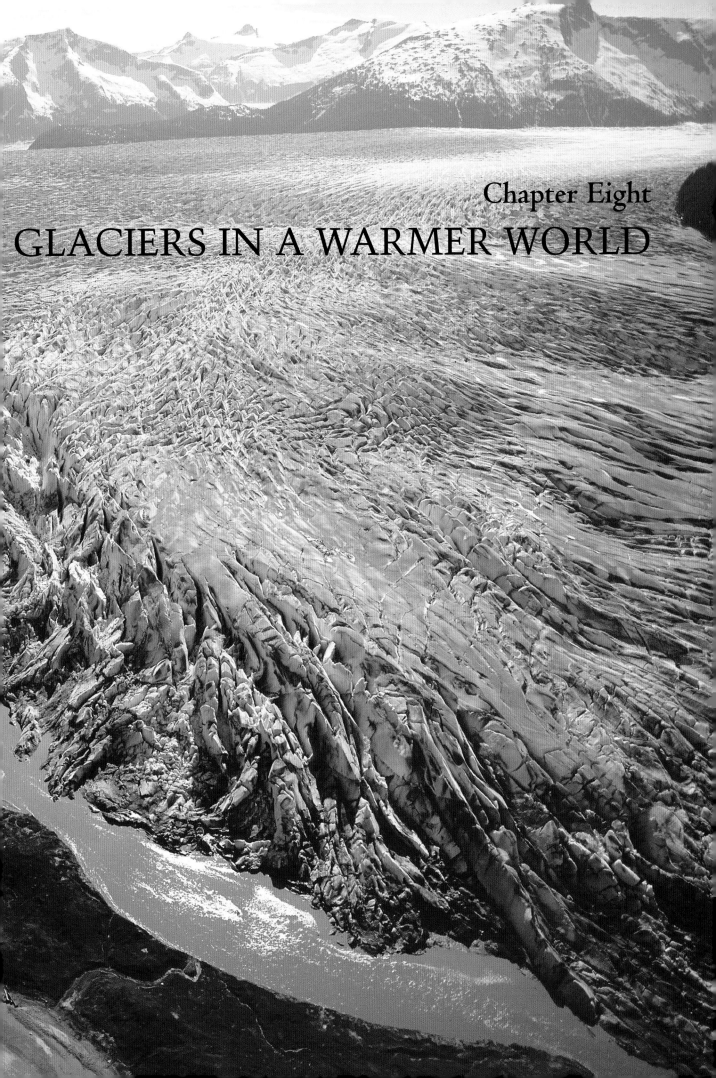

Chapter Eight

GLACIERS IN A WARMER WORLD

Dennis Trabant leaned against the snow shovel

and wiped his brow. He looked far into an intensely blue sky, trying to spot a plane flying somewhere overhead. The sound of its motor echoed everywhere in this upper basin of Wolverine Glacier. There: white with blue trim, high-wing, skis. The Piper PA-12 carved an arc around the basin's headwall, leveled its wings, and shed elevation as it sped down-glacier. At the bottom, it turned and came back to repeat this track.

We were on a smooth surface of the upper glacier. The Piper returned, this time settling softly onto the snow nearby. "Heck of a nice day," Keith Echelmeyer observed as he stepped out of his plane. We didn't disagree. His assistant grinned as Keith described how they had been stuck in surprisingly soft snow over on the Harding Icefield earlier in the day. Somehow, they had coaxed the Piper's skis up on top of the snow and were then able to take off. All in a day's work for Keith. He had known that Dennis and Rod March would be up here on the Wolverine, and thought he'd drop in for a visit. Alaska may be huge but the glaciology community is small and tightly knit.

Keith showed us his rig: a bush plane bristling with shock-mounted computers, dual frequency kinematic GPS receivers, and a gyroscopically oriented laser altimeter. With this gear, Echelmeyer has been drawing contours of glacier surfaces throughout Alaska. The data is startlingly precise—thousands of points sampled along a line running down-glacier,

each point with a surface elevation accurate to within a few inches. Keith thought he had flown profiles of about a hundred glaciers so far.

Dennis and Rod have monitored Wolverine and Gulkana glaciers for decades, carefully accumulating an uninterrupted stream of information. Other scientists peruse satellite images and massage immense amounts of data about glaciers around the world. Their modern methods may be faster but lack the immediacy of working out on the ice. Keith's efforts lie somewhere in between—real information about real glaciers throughout Alaska. He knows ice from Alaska to Greenland to Antarctica and everywhere in between. After the plane departed, Dennis turned and said, "When Keith is out here, he moves like a wolverine."

ALASKA IS CHANGING. Barrow has mosquitoes now. Houses in Fairbanks need to be jacked up as their previously frozen foundations thaw. Increasingly, the perma is leaking out of permafrost. An Inuit tribe is struggling to come up with a new word for "thunder," a sound they never heard in the past. The Bering Sea ice margin is breaking up earlier each year, stripping away the platform upon which walrus and polar bear have always survived during springtime. Average temperatures throughout Alaska have risen five degrees in summer and ten degrees in winter since 1970. Always looking on the bright side, our federal government has pointed out that this could facilitate shipping in ice-free Arctic seas, and offer a longer growing season for the northland. Just what we need: even bigger cabbages in the Matanuska Valley.

Glaciers, particularly valley glaciers in a maritime environment, are the true canaries in this coal mine. Equilibrium line elevations are climbing. Precipitation patterns are changing.

During the 1990s, Keith Echelmeyer and his colleagues at the Geophysical Institute in Fairbanks profiled sixty-seven glaciers throughout Alaska, representing 20 percent of all ice in the state. They compared their airborne data to traditional topographic maps first made in the 1950s. Echelmeyer found that the surveyed glaciers had thinned by an average of 20 inches each year from the 1950s to the mid-1990s. The glaciers had lost 12.5 cubic miles of glacially stored water annually, which translates to a 0.005-inch rise of sea level. But when Echelmeyer reprofiled twenty-eight of these same glaciers in 2000 and 2001, he found that the annual rate of thinning had increased to 71 inches. And from the mid-1990s through 2001, these glaciers had lost 23 cubic miles of water each year, raising sea level around the world by yearly increments of 0.01 inches.

Who cares about one one-hundredth of an inch? Perhaps the 100 million people who currently live within about three feet of sea level care. If all Alaskan glaciers were to melt tomorrow (which is quite unlikely), sea level would rise one foot, barely covering those lowlanders' ankles. But perhaps these people would also care about the accelerating rate at which other glaciers in the world are melting. Each year's increment of glacial wasting and rising sea level is added to the total of all prior years. It turns out that maritime glaciers (be they in Alaska or Scandinavia or Iceland or the Andes) are spectacularly sensitive to a warming atmosphere, much more so than the great ice fields of Greenland and Antarctica. Alaska's current contribution to sea level change is about twice that of Greenland, even though there is significantly less ice here. And then there are unanswered questions about the stability of ice in West Antarctica—were it to melt, sea levels would rise by 26 feet.

◄ *Partially submerged terminal moraine left by the retreat of Beloit Glacier, Blackstone Bay, Chugach National Forest.*

The world's atmosphere has warmed a little more than one degree Fahrenheit in the last fifty years. The effects on glaciers are more complicated than a simple increase of ablation. The distribution of heat around the globe is handled about

equally by oceans and air. As oceans heat, evaporation increases locally and more precipitation is available in some locations. Consequently, even though they may be retreating, maritime glaciers could actually be through-putting more ice.

Why is the atmosphere warming? (This is a loaded question.) World temperatures have always naturally oscillated on many wavelengths. El Niño and La Niña events are quasi-periodic temperature changes of the equatorial Pacific Ocean, popping up every three to seven years. During El Niño years, southeastern Alaska tends to receive more precipitation. The Pacific Decadal Oscillation warms the

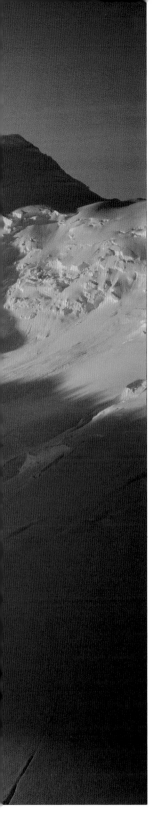

surface of the north-central Pacific Ocean almost a degree on a roughly twenty-year cycle. With this warming comes an intensification of that Aleutian Low. Climate change also occurs on a centuries-long scale. The Little Ice Age lasted almost three hundred years, and was preceded by Medieval Warm Epoch from AD 900 until about 1300. And of course, the ice ages have been cycling through on about a 100,000-year basis for the last two or three million years. So the Earth has a long track record of warming and cooling. Perhaps the current rise of atmospheric temperature is just another whim of nature.

Or perhaps not. In order to decide, it's worthwhile to tease out the various thermal inputs that force atmospheric temperatures one way or the other. Volcanoes can inject heat into oceans and the atmosphere. Subtle changes in the Earth's rotation and orbit around the sun can increase or decrease temperatures. So can changes in the atmosphere's concentration of carbon dioxide, methane, nitrous oxide, and chlorofluorocarbons—the greenhouse gases that tend to trap solar energy. You also must find a reliable climate record that is long enough to meaningfully examine. Records of thermometer readings extend back into the nineteenth century. Extrapolation further back in time requires proxy records using tools such as tree-ring analysis and $\delta^{18}O$ measurements in ice cores. Climatologists have taken temperature records in one hand and thermal inputs in the other, and compared them.

It turns out that volcanoes and orbital perturbations account for most observed temperature variation through about 1955. By the second half of the twentieth century, however, temperatures began to rise faster than could be accounted for by natural inputs alone. Beginning about mid-century, greenhouse gases caused more and more of the atmosphere's warming. Indeed, carbon dioxide has steadily increased in the atmosphere as industrial societies burn more fossil fuels. Using a "business as usual" model of continued greenhouse gas production, we see that worldwide atmospheric temperatures could easily climb five or more degrees Fahrenheit above a 1950s baseline by the end of this twenty-first century. And poleward extrapolation of average temperatures would likely be higher still.

What does the future hold for glaciers in Alaska? They've been thinning for two hundred years, first at the end of the Little Ice Age, and more recently with the onset of anthropogenic global warming. As the atmosphere continues to heat up, Alaska's glaciers will withdraw into higher and smaller niches within the mountains. This happened five thousand years ago when the Earth was hotter than we've managed to make it so far. Glaciers receded then, and came back after temperatures fell. There is little doubt that glaciers will continue to shrink. And they will likely return sometime in the future when the world cools again. The more personally urgent question is: Will our species be around to see such a cycle come full circle?

I looked down on Seward late one July afternoon: Sunlight streamed along the Resurrection River, rimming the trees near town in gold. Motor homes, reduced by my altitude to tiny silent toys, bobbed along the highway from Anchorage. A fogbank slowly crept up from the Gulf of Alaska, trying to cover Humpy Cove. There would probably be time to return before that fog swallowed Seward's airport. You pays your money and you takes your chances. I tugged on the reins of my Cessna 180, turning its head westward. Passing Paradise Creek, Exit Glacier provided a quick ramp into a wilderness of ice.

We sailed over crevasses, rattled over icefalls, and drifted onto the smooth back of the Harding Icefield that stretched away 50 miles to the southwest. The rolling pastures of snow glowed from within, a rich yellow honey running downhill where the sun still shone. Light ricocheted around cirques and peeled past peaks, pastel pinks fading into soft blue shadows. Isolated mountaintops—nunataks—pierced the icefield, reflecting red in the distance. Dozens of glaciers spilled away from this icefield in all directions, plummeting through their respective valleys toward the sea. Icebergs clogged Holgate Arm and jostled for position in the lake at the foot of Bear Glacier. Here, spread all around me, was ice in every imaginable form. I scarcely breathed, scarcely believed I was flying, so beautiful was this view.

➤ *Bear Glacier, Kenai Fjords National Park.*

Suggested Reading

Alley, Richard B., 2000, *The Two-Mile Time Machine*; Princeton University Press, 229 pp.

Bailey, Ronald H., 1982, *Glacier*; Time-Life Books, 176 pp.

Bennett, Matthew R. and Glasser, Neil F., 1996, *Glacial Geology*; Wiley and Sons, 364 pp.

Gordon, John, 2001, *Glaciers*; Voyageur Press, 72 pp.

Krimmel, Robert M. and Meier, Mark F., 1989, "Glaciers and Glaciology of Alaska"; in *Glacial Geology and Geomorphology of North America*, American Geophysical Union, p. T301/1-59.

Molnia, Bruce, 2001, *Glaciers of Alaska*, Alaska Geographic, 112 pp.

Muir, John, 1915, *Travels in Alaska*, Houghton Mifflin, 327 pp.

Post, Austin and LaChapelle, Edward R., 1971/2000, *Glacier Ice*; University of Washington Press, 145 pp.

Winkler, Gary R., 1999, "A Geologic Guide to Wrangell-Saint Elias National Park and Preserve, Alaska"; USGS Professional Paper 1616, 166 pp.

Acknowledgments

The US Geological Survey was an inspiration throughout this project. Dennis Trabant and Rod March introduced me to their wonderful world of glaciers and did a fine job of keeping me out of the crevasses. I learned a great deal from conversations with Austin Post, Robert Krimmel, Don Thomas, and Philip Hooge—all with the USGS. Elliott Spiker, Martha Garcia, and Bonnie Murchey at the Survey's Earth Surface Dynamics Program supported my work on and above the glaciers. Keith Echelmeyer (University of Alaska), Richard Alley (Penn State University), and Bernard Hallet (University of Washington) gracefully fielded my questions, no matter how simplistic.

I couldn't have asked for better kayaking companions than David Dinter and Tom Bean. Ed Peacock was the picture of patience and Stan Steck the perfect host when we were stuck at the Denali Strip under inclement weather. Thanks to Chris Smith and Meadow Brook, who took us into their home, sparing us four nights of camping in the Alaska Airline baggage carts at the Gustavus airport. A tip of the hat to mechanics Larry Gregg in Juneau, Cliff Belleau in Anchorage, and especially Jay Kerger and John Atterholt in Prescott, Arizona, for keeping my all-too-venerable plane aloft.

It must be obvious by now, but let me say it anyway: Working on this book was an incredible experience. Carole Thickstun knew I wouldn't pass up the opportunity when she first suggested I tackle the project. And I knew that I wouldn't pass up the opportunity to work with her designs or Lawrence Ormsby's illustrations. Thanks to Charley Money and the Alaska Natural History Association for turning me loose in Alaska, and to Susan Tasaki for rounding up the dangling participles when I finished.

Finally, I would particularly like to thank Jene Vredevoogd, who flew while I photographed throughout southern Alaska. Jene's easy humor, good sense, and great enthusiasm were essential ingredients in brewing up this book. And as always, Rose Houk inspires every picture I frame and word I write. My thanks to you all.